(法)苏菲·巴尔波 / 编著

夏国祥 / 中译
(美)柯尔斯顿·薛帕尔德 / 英译

SPONGE CITY
Water Resource Management

海绵城市

广西师范大学出版社
·桂林·

images
Publishing

图书在版编目(CIP)数据

海绵城市:汉英对照/(法)巴尔波 著;(美)薛帕尔德 英译;夏国祥 中译.—桂林:广西师范大学出版社,2015.9(2016.4 重印)
ISBN 978-7-5495-7124-6

Ⅰ.①海… Ⅱ.①巴… ②薛… ③夏… Ⅲ.①城市规划-空间规划-汉、英 Ⅳ.①TU984.11

中国版本图书馆 CIP 数据核字(2015)第 202503 号

出 品 人:刘广汉
责任编辑:肖 莉 夏永为
版式设计:张 晴
广西师范大学出版社出版发行
(广西桂林市中华路22号　邮政编码:541001)
(网址:http://www.bbtpress.com)
出版人:何林夏
全国新华书店经销
销售热线:021-31260822-882/883
上海利丰雅高印刷有限公司印刷
(上海庆达路106号　邮政编码:201200)
开本:635mm×965mm　1/8
印张:33　　　　　字数:100千字
2015年9月第1版　2016年4月第2次印刷
定价:268.00元

如发现印装质量问题,影响阅读,请与印刷单位联系调换。

FOREWORD
序言

Sophie Barbaux
苏菲·巴尔波

In the 20th century, city development was done on a deliberately invisible – or nearly invisible – rainwater-management model. This water passes from the rooftop gutters and drain spouts, flows the length of street gutters and continues to stream or pour into sewers dug into an impermeable soil, to join with multiple underground networks.

The development of new urban zones in cities and rural areas and the expansion of related ceaselessly growing infrastructures have greatly increased impermeable, macadamized, hard, dry and inhospitable surfaces – considered irreversible – in the process of connecting new areas to means of transportation, thus reducing the porous zones that naturally absorb rainwater.
The water cycle has thus been modified, with increased peak flows generated by thunderstorms and heavy rains, clogging collection networks and increasing the runoff volume that flows into the waterways, generating leached pollution. This organization of the urban hydraulic networks has shown its limits and inability to contain the flow it is required to manage. As a result, frequent floods and the ensuing concentration of contaminants have serious human, economic and ecological consequences.

From the late 20th century, the growing awareness of the ecological problems generated in this field, on the land and greater urban areas, has triggered awareness of the need for better control of the sanitation and utilization of rainwater. Rainwater management has therefore become one of the major objectives of the sustainable development approach, as exemplified by France's HQE (Haute Qualite Environnementale—High Environmental Quality). This standard for green building, which is starting to develop progressively around the globe, is particularly vital given that cities and metropolises are predicted to accommodate 70 to 80 percent of the world population by 2050.

在20世纪，城市发展中的雨水管理模式是以十分隐蔽的、甚至几乎无法看见的方式进行运作的。在这种模式中，雨水通过屋顶的排水道和垂直排水管来到地面上，然后沿着街道的排水沟或各种道路边缘沟渠继续流动，或者通过路面的排水孔进入不透水的地下水道，最终汇集到多层面的地下排水与储水系统中。

城市和农村地区的发展，以及因此而不断增长的相关基础设施，对于地表不透水地层面积的增加和透水地层面积的减少影响极大，然而在自然条件下，只有透水地层才能吸收地表雨水。为了在新开发地区建设交通网络，很多土地被铺上沥青、覆盖上硬质铺面，形成干燥和不友善的空间，而这些过程往往被视为是不可逆的。水的自然循环方式因此被改变，暴风雨造成的洪峰不断地增加，大雨使雨水排放系统陷于瘫痪，暴增的径流溢出水道，四处流淌，进而造成灾害与污染。既有的城市水网已经显示出其局限性，所要面对的水流处理量已经超过了它们的能力范围。频繁的洪泛和随之而来的水体污染物浓度增加，对人类健康、经济生产，以及生态环境产生严重的危害。

在20世纪末期，由于对土地和更广大范围的城市环境生态问题有了更深入的理解，人们清楚地意识到进一步加强雨水净化和雨水利用的必要性。雨水管理也因此成为可持续发展建设（例如法国"高环境质量认证"，即HQE）必须达成的主要目标之一，而此课题已经逐渐在全球范围内取得了很大的进步。如果未来全球城市和特大城市将按专家预期的那样容纳世界70%—80%的人口的话，这个可持续发展的课题更是至关重要。

许多建筑师在项目设计中积极将此问题纳入考量，尤其通过屋顶花园和屋顶平台水池的设置，来对从各建筑物汇集

Architects participate in this greater awareness of the problem, notably via the development of rooftop gardens and rooftop-terrace reservoirs permitting the recovery and use of recyclable water collected on different buildings. But it is the landscapers, in collaboration with urban planners, to whom falls the important mission of entirely rethinking existing or newly created developments and of overhauling rainwater management to meet the ever-expanding needs of infrastructure. This requires the abandonment of the 'all pipe system' and its replacement with alternative or essential compensatory solutions.

In order to respect the water cycle, it is necessary to take into consideration the affected site, as well as the areas downstream and upstream so that problems are not just moved further along. It is important to consider permeability as a support for the project, as it participates in the development of very diverse landscaping aesthetics, adapted to each context, whether a new creation or a rehabilitation. And in order to obtain permeability, It is necessary to use multiple techniques, renewing past, temporarily forgotten, practices as well as adapting innovative technologies in order to maintain the water within the affected area so that it can be filtered, used and even magnified. The goal becomes to reduce the wastewater reaching the old sewer system to the smallest volume possible.

The first developments were created to limit or even prevent floods. In these storm basins, an enormous network fed a more or less porous zone, bordered with fences, voluntarily maintained to limit the spontaneous invasions of weeds. Today, these works have multiple technical functions, but also serve to support the vegetation of the sites, to enrich the biodiversity and to integrate it into a more global landscaping project. Thus biowales, ditches of Caux, lagoons, as well as sedimentation tanks, create specific plant environments

而来的雨水进行回收和利用。然而重新整体性地思考既有设施的使用以及为其进行更新改造的重大任务，往往落到景观设计师与城市规划师的身上。这种重新思考导致了对"全管道系统"的扬弃，并且为此提出不可或缺的补偿性替代解决方法。

为了使水流有效循环，必须对规划基地建设进行整体考量，甚至包括雨水径流的上游和下游区域，才能避免仅仅将问题转移到其他地方。将地面的透水性视为方案设计的基础因素，无论对新建项目还是改造项目而言都是非常重要的，这个透水性的处理也必须能够因地制宜，而创造出各种景观美感。为了达到此目标，多样技术手法的运用是不可或缺的，不论是与暂时被遗忘的古老工法结合，还是采用创新科技，都必须最大限度地保存基地内的水资源，使水体能够就地被过滤净化和回收利用，甚至被彰显和美化，而后才将最少量的水导入早期建立的管道系统中。

人们都记得最初为了限制甚至预防洪涝灾害发生而设置的雨水管理设施，美其名称之为暴雨池（storm basin），这是一种由或多或少具有吸收水体能力的地块而构成的巨大网络系统，这些地块的周边建有围篱，以防止那些不请自来的植物侵入。如今，此种暴雨池拥有了更多的技术功能，不仅提供当地的植物灌溉用水，也用于增强生物多样性，并将水池融入一个更具整体性的景观设计方案中。生态草沟、围篱式沟渠、净水泻湖和沉淀池等空间形成了独特的植物生长环境，并且对地表水体进行收集、储存、净化、疏导和优化，使之成为名副其实的城市资源。蓄水池转化为湖泊或水塘，溪流、沟渠或水道被重新开凿或融入环境之中，不仅能够灌溉各种植物空间，也创造出全新

and contribute to collecting, retaining, purifying, guiding, and optimising this surface water, permitting it to become a nourishing urban resource. The water retention basins become lakes or reflecting pools, a new geography of landscape. The pedestrian or vehicle thoroughfares, like the parking lots, cease to be impermeable surfaces and instead provide greenery, cropped and varied, even flowering, to the contingencies of the modern world. Wetlands are no longer dried but transformed into wet meadows, riparian banks, or aquatic gardens, where people can wander at leisure.

This book brings together a panorama of French creations in this domain, which, without being exhaustive, explores the richness and the diversity of the landscaping approach and the technical solutions developed, or even invented. On an equally urban and peri-urban scale, the ecosystems that are preserved or created, autonomous or connected with one another, have been put in perspective, showing the importance of the environmental phenomenon in progress. They gradually transform the city into an immense green sponge, an image taking up the term 'sponge city', which is now widely used in Asia. Its scope often exceeds just the management of rainwater and runoff.

In fact, the solutions carried out by landscapers also take into account the indispensable fight against the 'heat island' effect, resulting from the mineral quality of the cities and due to the extreme concentration of buildings. This heat effect creates inert soils and generates the small amount of evapotranspiration emitted by a too-rare vegetation, creating both discomfort and the degradation of air quality. These two realities of the current urban environment pose real problems for public health. Water, like air, is a natural heritage and is crucial to preserve!

的景观地貌。而人行道、车道以及停车场不再采用不渗水的地面设计，取而代之的是绿意盎然、经过修剪整理的草地，以及多样的、甚至种植花卉的地面，让现代世界变得更加多姿多彩。湿地也不再一味地被干燥化，而是被整治为湿草原、河岸或水上花园，让人们在其中悠闲漫步。

本书将法国在雨水与径流管理方面的规划设计做了一个全面性的项目汇集，也许并非百无一漏，却有效地探索了景观设计手法和技术解决方案的丰富性和多样性，不仅对既有成规进行了检视、改良与发展，更不乏独具创意的构思。这些项目突显出当今环境课题的重要性：无论是在城市本身或涵括郊区范围的尺度上，许多生态系统都逐一受到保护或重新建立，其中有些形成了独立自主的系统，有些则互相联结，产生依存关系。这些生态系统逐渐地将城市转变成具有调节和运用水资源能力的巨大绿色海绵——在亚洲地区经常被使用的"海绵城市"一词为此种雨水和径流的管理概念赋予了清楚的意象。此类规划往往超越了技术与功能的目的，在解决问题的同时也大幅度地提升了景观和环境的品质。

景观设计师们提出的水资源管理方法同样也将"城市热岛效应"视为整治目标。此热效应来自于城市建筑的过度集中与城市设施的硬质化，进而造成缺乏活力的地表，同时过于稀少的植物只能向空中蒸腾少量水分，不仅影响环境的舒适性也使空气质量每况愈下。当前的这两种城市环境现实对城市居民的公共健康造成了实际的问题。水，像空气一样，作为大自然母亲送给我们的礼物，需要认真予以保护。

CONTENTS
目 录

003　FOREWORD
　　　序言

Water-scapes Permeability
景观与水敏性

010　Western Salted Field
　　　西部盐沼

014　Chemin de l'Île Park
　　　岛屿小径公园

020　Park of the Seille
　　　塞勒河公园

024　Planches Island Park
　　　普朗诗岛屿
　　　溢洪道公园

030　Clos Allard Park
　　　阿拉尔园地

034　Gassets Brook
　　　格赛小溪

038　Wet Meadows and Source
　　　of the River Norges
　　　诺尔热河发源地
　　　与湿地草原

Eco-Neighborhood Gardens
社区生态公园

046　Cent Arpents ZAC
　　　颂阿尔邦开发区公共空间

052　Landscape of the Sycomore
　　　Eco-Neighborhood
　　　梧桐生态街区
　　　景观

058　Camille Claudel Eco-Neighborhood
　　　卡米耶·克洛岱尔生态街区

062　Bottière Chênaie Eco-Neighborhood
　　　波提耶尔 - 申内
　　　生态街区

074　La Morinais Neighborhood
　　　莫利奈街区

088　Croix Bonnet Neighborhood
　　　克鲁瓦·伯奈特街区

092　Champ-Fleuri Neighborhood
　　　查姆普 - 弗瑞街区

098　EuroNantes Gare/Malakoff
　　　Neighborhood
　　　欧洲南特火车站 /
　　　马拉科夫街区

106　Baudens Neighborhood
　　　鲍登斯街区

110　Gaubert Building Complex
　　　高贝尔综合小区

The Management of Rainwater in City
城市雨水资源管理

116　Haute Deûle River Banks
　　　上德勒河岸

126　Le Havre City Entrance
　　　勒阿弗尔城市入口

132　Lucie Aubrac Square
　　　露西·奥布莱克广场

| 136 | Three Rivers Walkway
三河林荫道

| 142 | Large Meadow
大草原

| 148 | New TGV Station of Belfort-Montbéliard
贝尔福 - 蒙贝利亚尔高速火车新车站

| 154 | Zénith Concert Hall Public Realm
天顶音乐厅室外空间

Rainwater Parks
雨水公园

| 162 | Prés-Devant Suburb
布莱 - 得旺市郊

| 168 | Mount Évrin Park
艾夫兰山公园

| 174 | Glonnières Park
格隆涅尔公园

| 178 | Park of the Savèze
拉萨维泽公园

| 184 | Docks Park
码头公园

| 190 | Billancourt Park
比扬古公园

| 196 | Waterside Park
滨水公园

| 202 | Ampère Large Garden
安佩尔大花园

| 208 | Neppert Gardens
内贝尔花园

| 212 | Garden of Giants
巨人花园

| 218 | Fairy Enclosure
仙子园圃

Public Rainwater Gardens & Green Spaces
公共雨水花园与绿化空间

| 228 | University Campus of Alençon
阿朗松大学校园

| 236 | Saint-Dizier High School
圣迪济耶中学

| 240 | Citis Business Park
西提斯科技园区公园

| 244 | Renault Technocentre Park
雷诺科技研发中心公园

| 250 | Odyssey 2000 Gardens
奥德赛 2000 花园

| 258 | Saint-Gobain Building Platform
圣戈班建材商场

| 262 | ANNEX
The Designers
附录
事务所和设计师

Waterscapes Permeability

景观
与水敏性

Mutabilis Paysage & Urbanisme

Western Salted Field
西部盐沼

Location：地点
La teste de Buch, France
Completion date：完工日期
2009
Area：面积
44 ha
Client：业主
Ville de la Teste de Buch
Contracted partner：合作事务所
Raphael Zumbiehl, Didier Gourmelon, Sogreah
Photo credits：图片版权
Hervé Abbadie

On the Atlantic Coast, across from the Arcachon Basin, the project restores and enhances the salt marshes (also called schorres) of the town of La Taste de Buch, left fallow and abandoned for decades. The water is both a goal in the research of the reopening on the sea from the urban fringe, a resource that finds support to restore natural environments, but also a form of partnership in the creation of the park.

Mutabilis has sought to make the 'invisible' by restoring the natural environment as if it had always been there. On the sea side, seawater was utilized to clear the site, and the control valve for accessing seawater was restored in such a manner as to regain the richness of the ecosystems of the salt marsh's atypical and very rich environment, prized by wading birds.

On the land side, the site receives the heavily polluted water of the Craste, a small river that crosses the urban environment and recuperates the city's runoff water before reaching the site and flowing into the sea… The installation of lagoon gardens at the entrance of the marsh salts have improved the quality of the water that nourished the environments and made it possible to structure the urban fringe of the project.

Today, it is a place to walk and have contact with the natural environment and the sea, which was almost forgotten as so much had become distant and hidden. The salt marshes have rediscovered an instrumental value in this space where water is the principal actor.

该项目位于大西洋海岸阿卡雄海湾对面的区域，试图恢复拉·塔斯特·德·布奇城外盐沼地（亦称"schorres"）的生态环境，并且提升基地的价值，重振其活力。该处土地已经被遗弃并荒芜了几十年。获得水源是研究重新开发城市周边的海水的主要目的，水资源是恢复生态环境的重要基础，也是在该处建设公园的重要结构元素。

穆塔彼利斯事务所首先致力于创造"看不见的事物"，建构丰富而多样性的生物环境，犹如它们一直以来便存在一般。借助位于海边的条件，海水被用来改造该处环境，海水淡化设施的修复使人们得以重入荒地，再度找回盐沼与海滨的丰富性。这种地域环境是极受水栖鸟禽所喜爱的。

由于基地与城市环境直接相邻，方案的目标之一也在于对卡斯特溪所汇集的污水进行净化，使其随后汇入大海。潟湖花园的设置同时也提供了对城市边缘进行整治的机会。方案将卡斯特溪的渠道化部分拆除，将天然面貌还给这个城市空间。

如今这个基地已然成为一个散步场所，也是人们与自然环境包括大海接触的媒介。那个曾经变得遥远和隐秘的大海一度几乎被人所遗忘。这些盐沼通过方案找回了它们存在的理由，而水是实现这一切的主要元素。

Competition schematic plan of the water circuit and the different environments | 不同类型环境中水循环图解

An atypical environment, rich with salt marshes, is restored through the reopening of the site to seawater	01	这个遍布盐沼的特殊环境借助海水的重新引入而获得恢复
Paths of discovery skim these fragile spaces in which seawater and the water from the lagoon gardens are a tool for the restoration of natural environments	02	带人深入基地探索的小径浮空架设在这个脆弱的环境地面上，在此，海水与泻湖花园的净化水使基地的自然生态环境得以重建
The valve that controls the opening and closing of the site to seawater is also a look-out point over the Bay of Arcachon	03	控制海水出入基地的水闸，从该处可以看到阿卡雄湾

011

The water that crosses the urban environment upstream of the site is filtered in the lagoon gardens at the site's entrance before being restored to the other environments and then finishing its path to the sea
The reopening of the salt marshes makes it possible to redefine and highlight the language of the village of oyster farmers
Different uses of the restored environments

04 流经基地上游城区的水流在项目入口处的泻湖花园中得到净化，被利用于一些自然生态环境的重新建构，最后流入大海
05-06 盐沼的重新开发使得与基地相邻、以牡蛎饲养场为生的村落得以获得新生，并展现地理特色
07-09 基地的生态环境经过复原之后，提供人们多种用途

Section: lagoon gardens filter and purify the runoff water of the Craste Stream, part of the contiguous urban fabric

剖面图：泻湖花园过滤并净化了地表径流水和卡斯特溪从其所穿越的城市街区携带来的污水

Chemin de l'île Park (Path of the Island Park) is located on the banks of the Seine at Nanterre. Its development, in a maltreated area, proposed an alliance between the city and nature, a wonderful biological machine, to create a lush and fertile environment.

Water, an educational and ecological tool, is the motor of the project. Drawn from the Seine by worm drive pump, it is purified, running through a series of aquatic gardens before being restored to the park for the creation of several prolific natural environments. Collected, directed, enhanced, water flows by gravity through a route where it feeds the environments that it crosses (watering the clearings, powering the turbine, feeding the pond and cisterns of family gardens, and so forth). It also helps to create displays: waterfalls and green wall, lily gardens, tussock gardens…

Volumes of soil displaced by the terracing were reused on site and made it possible to shape the landscape of the park and the path of the water. The leveling of the soil and its opening on the Seine make it possible for the park to be flooded in case of spikes in the water level. The park, thus, offers a supplementary surface for excess water, which limits its impact on inhabited areas. Finally, the water structures the elements of the program (canal, ponds, cafés, welcome house, family garden) and makes the emergence of new uses possible in this transformed landscape (picnicking, fishing, discovery of the flora and fauna, jogging, cycling, gardening).

Mutabilis Paysage & Urbanisme + Guillaume Geoffroy Dechaume

Chemin de l'Île Park
岛屿小径公园

Location | 地点
Nanterre, France
Completion date | 完工日期
2006
Area | 面积
14 ha
Client | 业主
EPADESA, Conseil Général des Hauts de Seine, Conseil Régional d'Ile de France, Ville de Nanterre
Contracted partner | 合作事务所
Gilles Clément, Chemetov & Huidobro, Atelier Cépage, Site & Concept, Mizrahi
Photo credits | 图片版权
EPASA (n°01), Hervé Abbadie (n°02–03, 05–13), Marianne Feraille (n°04)

岛屿小径公园位于塞纳河畔的楠泰尔。该公园处在一个不受待见的地区，但仍旧取得了相当程度的发展。公园方提出一个观点：城市要与强大的自然界结盟，创造一个郁郁葱葱的回报丰厚的环境。

水，作为一种能为青少年提供教育、调控自然环境的资源，是本项目的发动机。纯净的水被不断地从塞纳河中泵出，经过一系列水上花园，被输送进公园，用于几处繁茂植物带的浇灌。水不断地被收集、引导、积蓄，通过重力流动，为它所流经的环境中的事物提供用途（浇灌空地、驱动涡轮，为池塘和私人花园提供水源等）。水也用于创建好看的景观：瀑布和绿墙、百合花园、草丛花园……

大量土地被改造成梯田，改变了公园的景观风貌和水流的路向。土地的平整和在塞纳河岸开荒，使得在出现洪峰的情况下，公园有可能被洪水淹到。在本设计中，公园因此对过量的水进行了额外的处理手段，以限制洪水对居民区造成影响。最后，在这个不断处于动态中的项目中，凡是涉及水的元素都被赋予新的任务，以便在满足景观设计需求的同时，开发出其他用途（比如供人野餐、钓鱼、探索科学知识、慢跑、骑骑自行车和做园艺）。

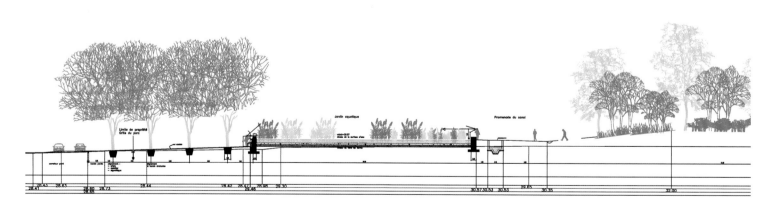

Section: the water is guided, channeled and exploited within the framework of the implementation of the environments
剖面图：在重新建构各种生态环境的过程中，水被引导、输入和利用

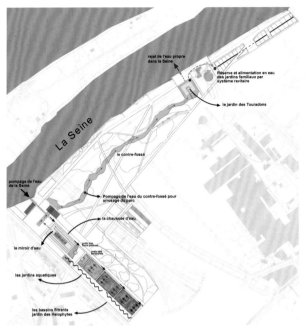
Schematic plan of the water circuit | 水循环图解

The framework of the park is developed along the Seine and offers a complementary area for the expansion of water in case of rising waters
Ponds planted with vegetation, conceived in collaboration with Gilles Clément. They filter the water that serves the creation of natural environments in the park

01 沿塞纳河岸建设的公园设施在洪峰期间为增加的水流量提供了一个额外的去处
02 与景观设计师季勒·克雷蒙合作构思的一系列水生花园，经这些过滤池塘净化后的水被利用来创造公园里的自然生态环境

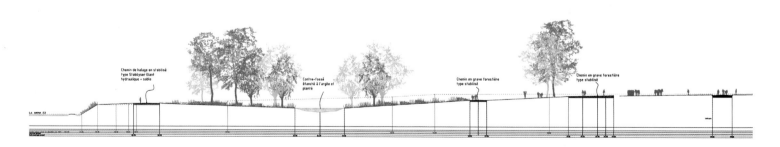

Section:
the side ditch flanking the Seine offers a supplementary surface for the rises in water level of the river

剖面图：
塞纳河岸的辅助水渠，在河流水位上涨时为增加的水量提供去处

Archimedean screw pump brings the water of the Seine to the highest filtering pond; the water course structures the open spaces of the park	03 螺旋升水泵结构将塞纳河水输入到一系列过滤池塘，而这些池塘的配置则组织了公园的主要空间
The side path is both an environment suitable for many uses and for biodiversity, as well as a water reservoir for the needs of the park	04 沿塞纳河岸设置的旁道水渠不仅为公园用水提供了储存空间，也是具有多种用途的场所，并且有利于发展生物多样化
Water accompanies visitors wherever they are	05 无论在哪里，水始终伴随着参观者的脚步
A new nature finds its place in the park: the management of water is the engine of the created atmospheres	06 公园呈现出一种崭新的自然面貌：水资源的妥善管理与运用促使了这些新生态环境的形成
The filtering ponds structure the urban part of the park	07 过滤池塘为公园的城市化空间建立了格局
Around an useful and educational route, water is exploited in all its forms	08 在这条具有实用性和教学性的路径上，水的利用以各种形式呈现出来

The horsetail pond collects the water purified in other ponds planted with vegetation and in the park surfaces	09	在公园地表和其他水生花园得到净化的水体汇入马尾草池塘
The island of the horsetail pond puts visitors in a cocoon of willows and isolates them from the noisiest parts of the park	10-11	马尾草池塘中的小岛让参观者仿佛置身于一个柳树编织成的茧壳中，浑然忘却公园其他部分的噪杂
Bio-indicators of the quality of the water, water lilies have been planted	12	漂浮于池塘中的睡莲不仅可供观赏，同时可作为水质量的自然指示物
The park house: the runoff water from the urban frink of the park and the filtered water of the Seine pour into the lateral ditch that flanks the Seine	13	公园之家：经过水生植物过滤的塞纳河水以及从公园边缘城市区域汇聚来的地表径流皆在此处流进设置于塞纳河一侧的水渠中

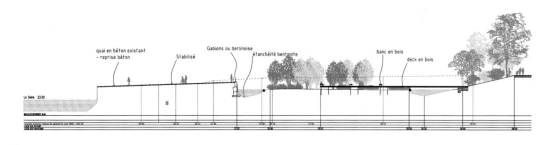

Section: | 剖面图：
garden of the Touradons, the last landscaped environment | 水流汇入塞纳河前最后一处
before the water returns to the Seine | 景观——图拉登花园

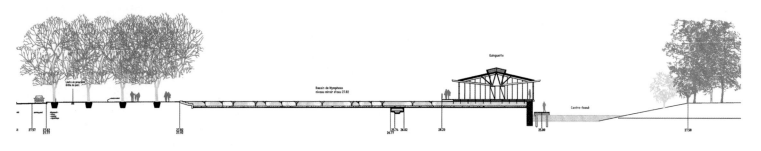

Section: | 剖面图：
the park house | 公园内的休闲建筑

Jacques Coulon

Park of the Seille
塞勒河公园

Location：地点
Metz, France
Completion date：完工日期
2002
Area：面积
20 ha
Client：业主
Ville de Metz
Contracted partner：合作事务所
Laure Planchais, Yves Adrien, Sinbio
Photo credits：图片版权
Jacques Coulon

The Park of the Seille program had two important orientations: urban and natural. At the time of the competition, the site was an industrial wasteland—a long band of sterile land flanking the Seille, a barely visible, steep-sided river. Jacques Coulon wanted to find the traces of the Seille in order to exhume its former geography. Upon rediscovery of traces of the ancient Seille, a branch was created with the purpose of increasing the volume of the banks and creating an inaccessible island where flora and fauna will develop more freely. In the city project, the park is the first phase of urbanization. It presages the birth of the Amphitheatre neighborhood.

Rainwater is managed by the park. Reed beds, lagoon, storm basin, and wet meadow regenerate the water and slow its natural course:

- Seasonal floods justify vast terracing. The 130 000 m³ of backfill that were extracted to allow the formation of the new branch and the softening of the banks are used to create a widening of the Seille's major bed in order to help regulate the river, several hundred meters from its confluence with the Moselle. This backfill will create the hills of the park.
- Visual opening of the area onto the riparian site. Today, the Seille is no longer a canal.
- Creation of lookout points from two 'hills'.
- Enhancement of the demarcation of the nature-culture line of the highest floodwaters, which becomes a long promenade that dominates the Seille and its banks.
- This line also implements phytoremediation strategies. It is in relationship with the stability of the new banks through vegetation engineering techniques, with the ecological water filter of the reed bed, the lagoon, the dry pool, and the wet meadows.

塞勒河公园设计方案有两个重要出发点：城市和自然。在参加设计竞赛阶段，项目基地仍是勉强可见、河岸陡峭的塞勒河旁的一片工业废弃地和一长条不毛之地。贾克·库隆为了研究塞勒河的古代地理情况，曾经探索过塞勒河的河道。根据这些研究，在塞勒河的古河道上，一条支流被挖掘出来，这样做的目的是增加河岸的高度，并以此造就一个隔绝在河中的小岛，以便让动物和植物种群在那里自由生长。在当地政府的规划中，在这里建造公园是当地城市化的第一阶段，它预示着一个崭新街区的诞生。

雨水将在公园中得到有效管理。芦苇地、人工湿地、暴风雨盆地和湿草甸可以在旱季释放出水流，并在需要控制洪水时减缓其流动速度：

- 季节性洪水需要通过整地工程来进行调节。挖掘塞勒河支流得到的130,000立方米土石，将被回填到适合的地方，塞勒河的干流将因此变得更加宽阔，增强其在跟摩泽尔河分流处以下几百米间的调节水位能力。回填的土石还将用于在公园内堆砌假山。
- 河岸区域，一目了然。今天，塞勒河不再是一条运河。
- 在"山"上造了两个瞭望点。
- 洪水最高水位线具有一种自然文化意义，它已经成为一条长长的标志线，统治着塞勒河岸。
- 这条线也已经成为环境植物修复的标准线。通过植物工程技术，并通过芦苇地、水泊、干燥池和湿地草甸的生态过滤，新河岸的稳定地位得到了确认。
- 项目的"文化"选择设计，最后落实在具体的"教育"花园功能上，

- The 'cultural' choice of the program, finally, is concretized by an 'educational' garden, where the perennials of the banks of the Seille are grouped by information panels.

Whether it was created directly as part of the project or whether it simply imposed underlying rules, the management of water is the prime mover. Any urban program will exploit it in planning a landscape authentic to and situated in its environment.

草木繁茂的塞勒河岸汇聚了众多植物，成为人们了解植物的展示馆。

无论是直接作为该项目的一部分，还是把水的管理看成只是强加的基本规则，水的管理都可以说是这个项目建设的原动机。确实，任何城市规划项目在进行真正的景观设计时，都会发挥水元素的作用，并将设计有机地融入对水的管理之中。

Doubling the size of the Seille by digging a new branch on the traces of its bed as it was in the 15th century

通过把塞勒河 15 世纪的一条旧支流重新挖掘出来，塞勒河的河床宽度变成了原来的两倍

01 Over the river banks, a few hardwood balconies
02 The Seille was a canal; today the opening of the newly floodable banks gives it back its status as a river

01 河岸上的硬木阳台
02 塞勒河本来是一条运河，今日新防洪河岸的开发使它成了一条真正的河流

A quay of long steps flanks the new branch of the Seille, it frames the little islands, accessible only to birds 03 沿着塞勒河新支流设置的阶梯状长条堤岸，新支流在河中心造成一个小岛，只有飞鸟可以直接登陆
The bankside walkways are all floodable 04 河岸的步道都具有吸收泛洪的功能
Just 10 minutes from the train station by foot, a very natural option 05 从火车站走过来只有十分钟路程，这里提供了一条接近自然的路径选择
The high paths, like a balcony over the river, mark the line of the highest water level 06 高处的散步道能够眺望河岸空间，同时也成为河流最高水位的标识线
The reed bed: the first pond of phytoremediation accommodate the rainwater of the new neighborhood of the amphitheatre 07 芦苇地：新近开发的剧场街区的第一个雨水净化池塘
A playing field with wooden steps 08 一旁设有木阶梯座位的游戏草地

HYL Hannetel & Yver

Planches Island Park
普朗诗岛屿溢洪道公园

Location：地点
Le Mans, France
Completion date：完工日期
2008
Area：面积
3 ha
Client：业主
Ville du Mans, Le Mans Métropole
Contracted partner：合作事务所
Arnaud Yver Architecte, Berim, ISL
Photo credits：图片版权
HYL Hannetel & Yver

In the center of Le Mans, an industrial island has become one with nature again… thanks to its new function as a 'hydraulic machine', serving as an overflow weir between the River Sarthe and its canal during high waters.

As much as possible, HYL had to unify the space, which was fragmented by the weir, three bridges, disparate banks, and a small housing operation. The plan was to turn the main part of the island into a lookout point through the use of riprap, strengthened embankments, and brick supports. This high point becomes a kind of bastion creating a dialogue with the walls of the housing estate, while the weir forgets its technical vocation and becomes the setting for a grassy amphitheatre and a long ramp where people come to sunbathe and picnic. On the canal side, the slope of the bank has been left naturally softened to allow people to look down to the water without any guardrail.

A small river citadel made of brick, stone, and earth, Planches Island Park restores to the confluence its geographical unity and provides new, attractive views for the housing estate. With its bridges making it far more accessible, the island has become the umbilicus of several neighborhoods coexisting harmoniously in the unifying landscape of the Sarthe.

位于勒芒市正中心位置的一个旧工业岛屿因为获得一个全新功能而得以重新转变成一片自然之地。这个新功能即是成为一个"水利调节设施"，在涨水期作为萨尔特河以及其运河之间的溢水系统。

方案必须尽可能地把被溢洪道分割成的零落空间、三座桥、不协调的河岸以及一个小型房地产项目等元素结合起来。设计师决定通过石砌护坡、加固路堤、砖砌防护墙等设施，来将岛屿的主要空间规划为观景平台。当溢洪道的技术功能被暂置一旁的时候，这个制高点便犹如与城墙对话的堡垒，它也变成在一个绿地剧场和一条长散步坡道上所凝望观赏的舞台景观。剧场和散步道同时化身为日光浴场，并接纳来草地上野餐的人们。靠近运河的一边，河岸的缓坡以天然状态呈现，并柔和地过渡到水边，人们在此无需任何围栏的护卫。

普朗诗岛屿公园犹如一座由砖块、石头和黏土砌成的河畔小城堡，它重新为这个河流汇流地带来地形上的统合，并且在城市中心缔造出新的景观视野。方案实施之后，岛屿上的桥梁拉近了它与人们的距离，它也成为那些和谐融入萨尔特河景观中的几个街区的"中心点"。

Master plan ┆ 总体规划图

Amphitheatre and spillway　01　剧场和溢洪道
Flooding of the spillway　02　溢洪道上的洪水

The functioning of the spillway during rises in the water level   03   洪峰时期溢洪道的运作
Amphitheatre seats   04   剧场座位
Green terraces   05–06   绿色休憩平台

027

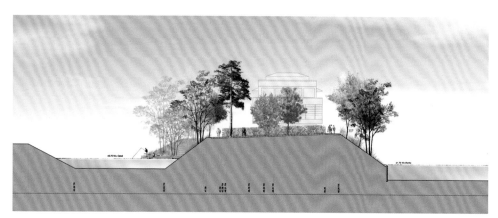

Transverse section of the children's island ┊ 儿童游乐岛剖面图

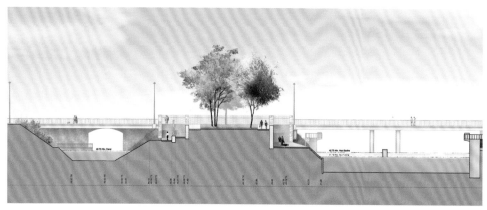

Transverse section at one of the park entrances ┊ 公园入口剖面图

Descent to the spillway 07 沿溢洪道而下即景
Terraced quay, the Sarthe side 08 萨尔特畔的堤岸平台
The large promenade flanking the canal 09 运河一侧的宽阔散步道

TN Plus

Clos Allard Park
阿拉尔园地

Location：地点
Caudebec-les-Elbeuf, France
Completion date：完工日期
2008
Area：面积
6 ha
Client：业主
Communauté d'agglomération d'Elbeuf Boucle de Seine
Contracted partner：合作事务所
TERAO, Phytorestore
Photo credits：图片版权
TN Plus

This project offered an opportunity to restore water to a central place in the composition of a park. More than just forging a link with the Seine, it was about restoring fullness to this landscape, and to value and use the relationship with water in all its forms, including drain rainwater. The challenge was to combine a tradition of the naturalistic park (in this vast natural setting of the valley of the Seine) with a contemporary image, a modern plan reconciling leisure uses and a landscape ambition that does not ignore the urban dimension of the project.

On rainy days an ephemeral landscape appears here, the watery reflections creating an ethereal space floating between the elements. This striking and poetic image leads us to consider the presence of water or its evocation as an additional element to describe these spaces. The project uses this echo of the Seine in proximity to the business park to render the setting more attractive. This work with water is closely connected with an HQE (High Environmental Quality) approach to the collection and treatment of surface water. Why not imagine the ground as a surface that captures, retains and directs the rainwater runoff? A project that controls the flow of water by stocking it before filtering it responds at the same time to an insistence on high-level environmental standards.

This project imagines and designs a controlled landscape that does hinder the reading of the bank, as a unitary space, with boundaries and contours that extend much farther; a landscape that magnifies the water circuit, which holds it in very shallow pools so as to make its evaporation visible, while cleaning it at the same time.

这个整治项目重新赋予了水元素在公园布局里的中心位置。在此方案中，水除了是唯一跟塞纳河产生联系的元素之外，其力量更在整体景观规划中获得全面的发挥，以各种形式来强化其价值与用途，包括雨水的排导。设计的重点在于将一个自然公园（地处塞纳河河谷的自然环境中）的传统本质和当代的意象相结合，透过一个具有现代感的规划，使"休闲"用途与景观企图相行不悖，同时不忘方案本身的都市尺度。雨天显现的短暂景观在此方案中透过水的反影塑造出一个在公园不同元素之间浮动的缥缈空间。这个令人屏息的诗意景象，引导着我们以水的呈现或是水的联想作为美化空间的附加元素。此外，本方案运用了邻近商业区的塞纳河所唤起的意象，让环境变得更具有吸引力。

这项水研究工作与"高质量环境认证"（HQE）关系密切，因此地表水的回收与处理也成为方案关注的重点之一。大雨过后，一处处的水洼表面片段地映照出了许多小片天空。在一片天空和另一片天空之间呈现了暂时性的景观，给人奇异的感觉。这个吸引人的景象，让我们一反过往的习惯，重新思考处理雨水的问题。为什么不将地面想象成一个能够聚集、引导和留住水流的表面？用这种思路设计的方案，能够先将雨水储存，之后再让它渗入地面，因而能够掌控地表的水流量，也回应了高环境质量认证的要求。

这个方案勾画出一种细致精算的景观，却毫不阻碍人们将整个水岸地带作为整体性空间的景观阅读，其界限和轮廓不断延续到远方。这是个礼赞水循环的景观：将水留在清浅的水池中，让水的蒸发明显可见，并同时对水进行净化。

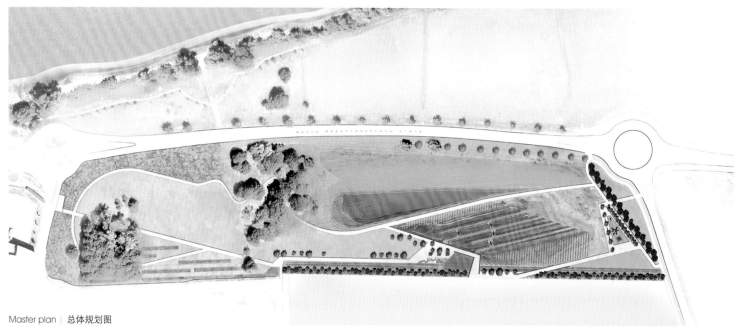

Master plan｜总体规划图

Reflecting pools 01 映象池
The large east-west ditch 02 东西走向的主要生态贮流沟
Playground 03 儿童游戏场

The large flowered lawn	04	花卉丛生的大草原
Climbable sculptures by the Simonnets	05	艺术家西蒙内设计的可攀爬雕塑
Bench on the lawn	06	草地上的长椅
The thicket of trees	07	树丛与草地

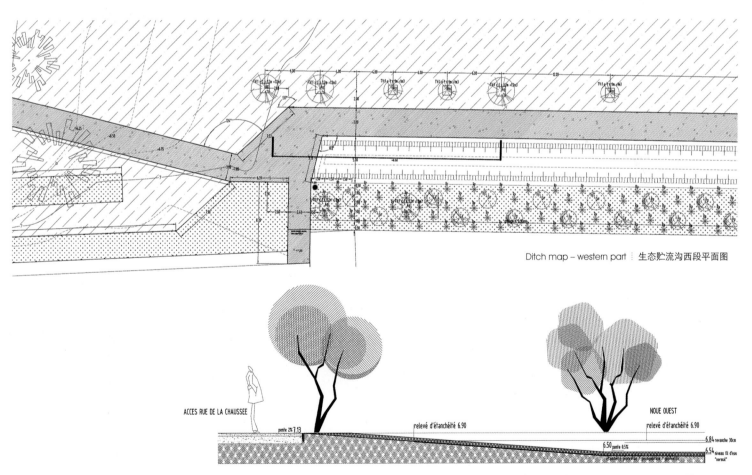

Ditch map – western part ｜ 生态贮流沟西段平面图

Section of the ditch – western part ｜ 生态贮流沟西段剖面图

Urbicus

Gassets Brook
格赛小溪

Location：地点
Serris, France
Completion date：完工日期
2001
Area：面积
3 ha
Client：业主
EPA Marne
Contracted partner：合作事务所
Tugec Ingéniérie
Photo credits：图片版权
Urbicus (n°01, 04–08), Éric Morency (n°02–03)

Gassets brook, a stream running down from the Briard plateau, structures the town of Serris. The development program envisaged lining the stream and channeling it underground to transform it into a wastewater facility beneath the public space. The main challenge of the project was to convince the local authority to keep the stream above ground and to enhance it through the creation of connected public spaces. The project therefore suggests an alternative stormwater management system and transforms the banks into a floodable walk.

The significant gradient between the stream and the public areas features a varied embankment of willow fagots and gabions. Several sequences create a journey from the 'natural' stream to its outfall into a large storm basin. The stream opens onto the town through a series of lookout points and holds sloping towards the water. Paths and flights of steps allow the public to go down to the brook and sit on its banks.

Riverside vegetation and wet meadows encourage a specific biodiversity, producing an effect of 'nature in the town'. Recently, however, over-intensive management of the vegetation has undercut the richness of the environment. This project shows how the issue of management is paramount. By drawing up a management plan it will be possible to follow and encourage its evolution over time.

格赛小溪是布利亚尔高原溪流的一部分，也是赛里镇的空间结构要素。改造项目最初计划在公共空间地下设置水流管道，将小溪整治成地下化的卫生工事。方案的最大挑战在于说服作为业主的公家整治机构让溪流维持露天状态，并建立一些与其相关的公共空间。因而方案建议以其他替代方式来进行雨水处理，同时将小溪沿岸改造成在洪泛时期可被淹没的散步平台。

小溪和各公共空间之间巨大的高差展现了多种河岸处理手法：斜草坡、柴笼或者石笼。从"自然状态"的溪流到注入防洪水池之间的整段排水口被细心处理成不同的景观片段。小溪以错落有致的观景露台和向水面倾斜的船坞来与城市产生对话，一些小径和台阶让人们可以穿过溪流或者在岸边停留休憩。

沿河的树林和湿地草场发展出独特的生物多样性，形成了"城市中的自然"。然而，最近一段时间，对植物的过度维护使生态环境变得贫瘠，于是方案也建立起一个适当的管理计划以顺应这个城市自然空间今后的发展。

The project permits the integration of the hydraulic system in the urban fabric | 项目允许对城区水管理网络系统进行整合

The brook is framed and enhanced by the landscaping　01　整治方案使这条小溪展现崭新面貌
Evolution of the neighborhood between 1998 and 2004　02-03　邻近地区在 1998 年和 2004 年间的发展

A plunging brook leading to a diversity of the profiles of river banks 04 项目的实施展示出多姿多彩的河岸
The little square like a balcony on the storm retention pond 05 暴雨蓄水池上仿佛阳台的小广场
A landscaping project promoting the access to water 06 整治方案为人们提供了更多的亲水空间
Details of the landscaping design 07-08 方案设计细部

Agence Territoires

Wet Meadows and Source of the River Norges
诺尔热河发源地与湿地草原

Location：地点
Norges-la-Ville, France
Completion date：完工日期
2013
Area：面积
0.29 ha
Client：业主
Commune de Norges-la-Ville
Photo credits：图片版权
Nicolas Watelfaugle

The town of Norges-la-Ville, to the north of Dijon and at the foot of the Langres plateau, is divided into two villages on either side of the Norges. This river takes its source from the foot of Norges-la-Ville. Frequent floods of the river occasionally cover the landscape of the natural banks and deprive the locals of its wild beauty.

For this project, Territoires chose a radical and minimal device: a wooden path that is as smooth as a knife blade. It rests on wooden legs, without foundation, like simple stilts that carry the visitor above the wet meadow.

This path extends across both villages, crossing the river by a footbridge halfway along. The geometric contrast between the winding river and the taut lines of the path gives its visitors an astonishing sense of levitation, without ever touching the ground. As such, this landscape is rediscovered, served by a radical and uncompromising aesthetic, with minimal human intrusion.

Within the framework of the project, the flood plains were preserved: the bed of a small tributary of the Norges was re-drawn and once more accommodates an aquatic flora and fauna, as a ball field was simply leveled and planted with a meadow that can be covered during winter floods. The construction of a wooden path was thought to present minimal hinderance to the movement of the river so that it can always pour into the wet meadows that border this riverine landscape.

位于第戎北边、朗格勒高原脚下的诺尔热·拉·维勒村庄包含了两个沿着诺尔热河发展的聚落，诺尔热河的发源地便在此处。然而，河流经常性的泛滥使得居民无法享受此地的旷野景观之美。

大地景观事务所（Territoires）为解决此问题选择了一个既根本又极简的设计：一条细致平薄的木板步道。该步道为木脚结构，没有地基，好像一排高跷一样立在湿地草场上，供人们在上面通行。

这条步道延伸于两个聚落之间，在一半的位置借助一段木桥通过诺尔热河，在河面上与河流保持着时远时近的距离。河流的蜿蜒曲线和步道的笔直线条形成强烈对比，使在步道上散步的人们仿佛身处于奇特绝妙的悬浮状态中，感觉双脚并未碰到真实的地面。此方案借助贯彻全程的美学元素和极简的人为介入，使人们重新找回此处的大地景观之美。

在设计的规划中，平原地区的蓄洪功能得以开发：诺尔热河的一条小支流被重新修整利用，该支流一度曾被众多的水生植物和动物所占据；邻近河岸的一个球场被平整出来，改为草场，以便在冬季水位上升时，将河水引流到那里。木制的步道设计是为了尽可能小地给河流造成阻碍，以便河流总是可以顺畅地流入湿地草场。那些湿地草场就分布在步道周边，因植物种类的不同形成多样的景观。

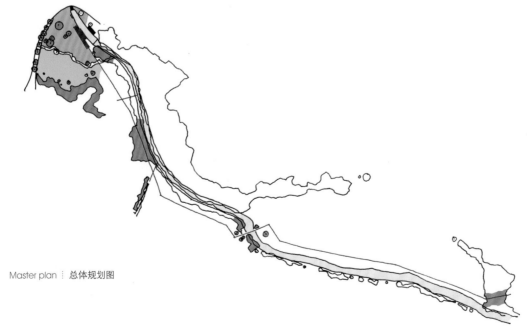

Master plan 总体规划图

The source of the Norges River and the ancient washing area 01-02 诺尔热河源头的古代洗衣区

The wooden walkway runs the length of the river, through the meadow and the ancient poplar grove

03-07 沿着河岸设置的木质步道穿过草原和昔日的杨树林

Constructed on pilings, the wooden walkway skims over the wet meadows and crosses the river

08-10 架高悬浮的木板步道穿梭于湿草地上并跨越河流

Eco-Neighborhood Gardens

社区生态公园

Florence Mercier Paysagiste

Cent Arpents ZAC
颂阿尔邦开发区公共空间

Location｜地点
Bussy-Saint-Georges, France
Completion date｜完工日期
2013
Area｜面积
57 ha
Client｜业主
Commune de Bussy-Saint-Georges
Contracted partner｜合作事务所
Tugec Ingénierie
Photo credits｜图片版权
FMP (n°01–07), FMP / Hervé Abadie (n°08), Epamarne / Éric Morency (n°09)

Based on the government agency EPAMARNE program, the development of the public spaces of the new neighborhood was carried out in successive steps, in parallel with those of the built parcels. It had to permit both roadway access to the future neighborhood as well as the management of surface rainwater. The hydraulic system is composed of transport apparatuses (ditches and channeling) and storage devices (ditches, pools, floodable spaces). Water pours into a last pool that makes the regulation of water leakage of the entire system possible.

One of the main ideas in the conception of these spaces was to reintroduce, at the heart of the new neighborhood, images of the countryside of Seine-et-Marne, through different landscaping motifs referring to it, while articulating contemporary images of the neighborhood in conjunction with the issues of biodiversity. The topic of stormwater treatment, with the topographic constraints that result from it, has led to a formal and aesthetic work at each of the spaces and the design of interior micro-landscapes in a quest to find a poetics of place. The system of collecting and slowing down rainwater was totally integrated into the conception of the private and public spaces. Each block of buildings had an obligation to manage the water through slightly sunken gardens. Runoff water from the roads (that don't have borders) is directed towards the open-air transport ditches that empty into storage pools, the design of which reconstructs a new landscaping look. A study of the topography and the interconnection of functions joins forces with a study of the vegetation, generating a receptacle for biodiversity.

在马恩省公共整治机构提出的规划基础上，与街坊地块的建筑工程同时间进行的新街区公共空间整治项目分阶段渐次完成。这个规划项目一方面为未来社区提供道路系统服务，另一方面可以收集地表的雨水——由不同传输装置（排水沟及运河）和储存装置（排水沟、水池、洪泛区）所组成的水力系统将汇集的雨水导入最后一个水池，以便有规律地泄洪。

空间规划与构思的主旨之一，就是利用取自塞纳-马恩省本身的各种景观画面在新街区的中心重新引入田园风光。地表雨水处理的主题促使设计要考虑地形条件，也在每个不同的空间尺度上诱发了一种形式上的造型研究与设计，内部微型景观的构思则参与了对场地的诗意的探寻。雨水收集和减缓雨水流动速度的系统，完全融入到私人和公共空间的发展规划思路中。每一座建筑物都有义务通过设置地势较低的花园来管理水资源。雨水从道路（没有边界）被导入露天运输渠道，再流入储存池。经过设计的存储池构成一种新的景观。地形的考察和功能的互联，以及如何种植植物，被综合考虑进一个具有生物多样性的包容空间。

为道路设计的全新景观也在此前提下被塑造出来，在经过调整的交通空间中，与汽车、行人和若干乡野景致片段交织在一起。马尔罗林荫道是一条"综合性"的道路（服务于行人、自行车及汽车），蜿蜒曲折地铺陈，并且连接着几个坐落在一系列形如群岛般平台

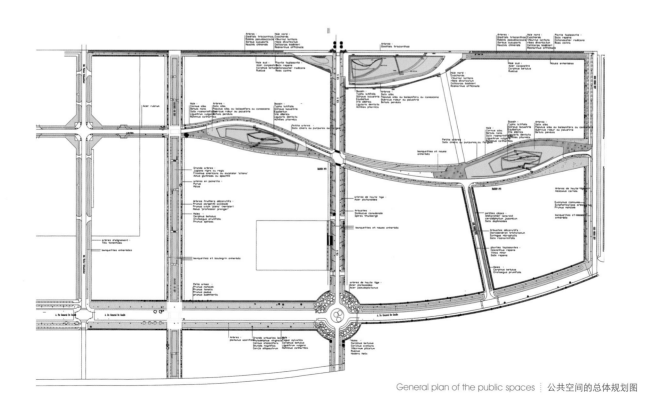

General plan of the public spaces｜公共空间的总体规划图

Entirely new landscapes for the roads were thus created, bringing together cars, pedestrians and bits of the countryside in a space with calmer traffic. Malraux Boulevard spreads out its meanders by integrating pools composed of a series of horizontal plateaus designed in archipelagos, in the center of the 'mixed' roadways (for pedestrians, bikes and cars). The J.-de-Tout, a planted boulevard accompanied by a lawn serving as a storage pool, crossed by canals, now has a view of the Genitoy Farm, while at the north, the space for water collection is transformed over the years, sometimes into a garden for river dwellers, sometimes into water pools.

上的水池；贾克·得·图散步道伴随着一片有水道穿梭的草坪，为人们带来朝向热尼陶农场的视野；北面汇集雨水的空间随着季节而变化，时而成为河边居民的花园，时而变成蓄满雨水的池塘。

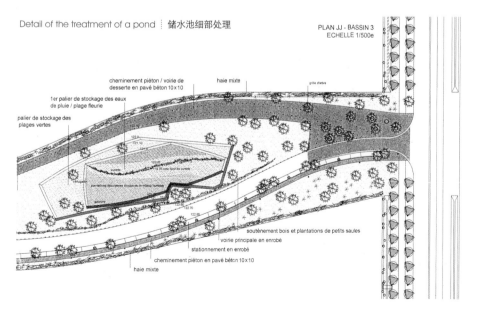

Detail of the treatment of a pond　储水池细部处理

The outline of the ponds recomposes　01　储水池的分布为项目建构出
the interior landscapes　　　　　　　　 崭新的内部景观

A view of the Genitoy: a succession of ponds bordered by two planted walkways 02-04 朝向热尼陶农场的景观视野：两旁设置了植物绿道的一系列储水池

Playing with the geometry and the vegetation of the ponds introduces a new poetic　　05–09　储水池的空间形式和繁育植物共同打造出一个诗意的空间

Florence Mercier Paysagiste

Landscape of the Sycomore Eco-Neighborhood
梧桐生态街区景观

Location：地点
Bussy-Saint-Georges, France
Completion date (part 1)：完工日期 (tranche 1)
2014
Area：面积
118 ha
Client：业主
EPAMARNE
Contracted partner：合作事务所
TGT & Associés, Tugec Ingénierie, Philippe Almon
Photo credits：图片版权
**FMP / Florence Mercier (n°01),
FMP / Antoine Duhamel (n°02–07),
Epamarne / Éric Morency (n°08–12)**

The new Sycamore project aims to create a diverse and sustainable eco-neighborhood that is integrated into the city and its history while meeting the challenges of the metropolis. Urban and landscaping frameworks were worked conjointly, linking the different programs together with the development of public spaces and the hydraulic system.

The creation of two parks, with a view towards the farm of the Génitoy, the interconnectedness of the landscaping spaces, and the project of an agricultural park all reinforce the dialogue between city and agriculture. The park of the Génitoy, created in 2013, makes the management of rainwater of the first part of the neighborhood possible, thanks to a large canal integrating into a sunken topography that permits the retention of surplus rains. The large lawn and the canal make up a system allowing for the overflow of water.

This park, besides its hydraulic function, accommodates numerous uses affiliated with urban culture and rural culture. A variety of spaces enliven the soft slopes of the park: sports fields, skate-park, amphitheatre, promenades, large lawn and a garden 'for the sharing of know-how' on the theme of growing and gardening.

The second park, called the 'Sycamore', opens onto the agricultural space. It will be composed of free and rustic lawns, punctuated with woods, and has a view of the horizon. The spaces of this park are closely linked to the management at the surface of the new neighborhood's rainwater. Large, naturally-shaped, pools will include hydrophyte vegetation. These pools will have a role in forming the promenade but will also be educational, focusing on the question of the enrichment of the biotope.

新梧桐生态景区项目的建设目的是创造一个多样化的、可持续发展的生态景区，同时在景区中融入所在城市及其历史文化，并能满足其面向大都市的建园目的。城区建筑和园林绿化工作采用相互协调策略，运用公共空间的发展和水元素系统将不同区域整合为一体。

生态区内有两个公园，面对着热尼陶农场，借助景观设计将常规公园和农业公园联系在一起，以此强化城市文化和农业文化之间的交流。热尼陶农场建造于2013年，对邻近地区的雨水管理做出了初步的贡献，这要感谢该公园内整合进的庞大管道，可以将洪水导入到地势较低能够容纳较多雨水的地理环境中。在那里，庞大的草场和管道系统共同构成了排洪系统。

公园除了具有水管理方面的功能外，还有许多与城市文化和乡村文化相关的用途。斜坡公园区设置有各种活动空间，共同活跃了公园的气氛，其中包括运动场、滑板公园、剧场、长廊、大草坪和一个有关成长和园艺的"知识共享"花园。

第二个公园，被称为"梧桐园"，是一个展示和参与农业文化活动的空间。这里包括自由生长、风格质朴的草坪，中间零星点缀着树林，在这里还可以看到广阔的地平线。这个公园与新邻居地表的雨水管理有着密切的关系。自然生成的巨大水池内有大量水生植被。这些水池有一定蓄洪作用，但也可以用于教学展示，主要呈现的是生物群落的丰富性问题。

General plan of landscape project | 景观总体规划图

In the middle of the park, the garden 'for the sharing of know-how' 01 公园中间的"知识共享"花园
The lines of the amphitheatre, where uses of the park are intensified 02 公园剧场周围的线条空间,其使用性相对密集

Ecological treatment of rainwater in public spaces | 公共空间中雨水的生态化管理

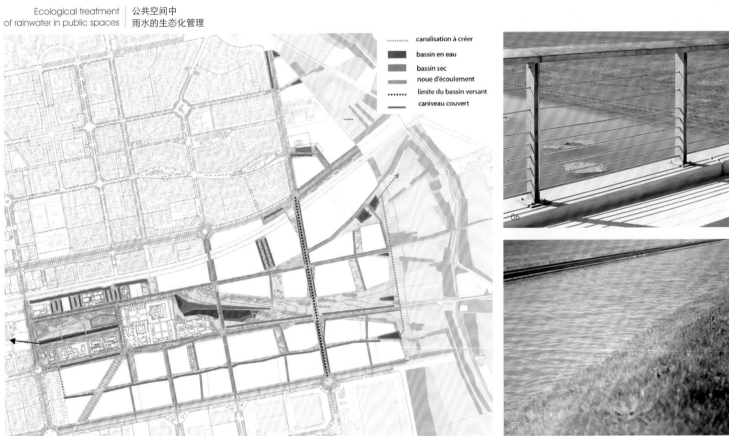

- - - - - canalisation à créer
▬ bassin en eau
▬ bassin sec
▬ noue d'écoulement
· · · · · limite du bassin versant
▬ caniveau couvert

The contours of the skate park　03–04　滑板场的造型
A canal with softly sloping banks in the middle　05–07　公园水利系统中的水渠
of the hydraulic system　　　　　　　　　　　　与缓坡草岸结合

An active border that accommodates multiple uses 08-09 具有多种用途的活跃边界
The garden 'for the sharing of know-how', a place of exchange on 10-12 "知识共享"花园成为一个
the theme of plants 以植物为交流主题的场所

Phytorestore / Thierry Jacquet

Camille Claudel Eco-Neighborhood
卡米耶·克洛岱尔生态街区

Location｜地点
Palaiseau, France
Completion date｜完工日期
2014
Area｜面积
12 ha
Client｜业主
Communauté d'Agglomération du Plateau du Saclay, Scientipole Aménagement
Contracted partner｜合作事务所
VIABE
Photo credits｜图片版权
Phytorestore / Thierry Jacquet

The landscaping project of the new sustainable urban neighborhood Camille Claudel, located on the Plateau of Saclay, plans housing for a population of about 4500 inhabitants. Through a functional mix (housing, schools, gym, and so forth), a durable and sustainable neighborhood will be created and integrated into its wooded environment through a voluntary eco-landscaping approach.

Eco-landscaping and water management over the whole neighborhood has been designed in a new experimental landscape framework.

Rainwater is managed opencast by:
- filtering equipment to collect and treat water from the roads, instead of the hydrocarbon separator;
- channels and retention basins built into the landscape that can store without leaking the rainwater from the heavy floods that happen every fifty years or so.

Part of the gray water from households is treated with compact organic filters and reused to water green spaces.

位于法国萨克雷平原上的卡米耶·克洛岱尔新区将包括住宅、学校、健身房、接待中心等功能混合区，容纳4,500位居民。这一空间得益于生态景观规划，将成为一个融入周围森林环境、富有生命力并具有可持续发展能力的生态小区。

该生态景观运用一种崭新的实验性景观结构，在景观规划的同时，对整个小区的水资源进行管理。

方案对雨水实施露天管理：

- 以过滤斜沟设备搜集并净化地表径流，替代油水分离系统；
- 将水渠及池塘融入景观规划整治，可以调蓄50年一遇的暴雨量，实现零排放。

一部分生活灰水通过紧凑的有机过滤系统净化后，被用以灌溉小区绿地。

Eco-neighborhood master plan｜生态街区总体规图

Parvis in front of the school　01　学校前院
Parvis from the side of the eco-neighborhood　02　邻接生态街区的学校前院

Large rainwater ditches　03-04　主要生态贮流沟
Rainwater ditches in the middle of city blocks　05　街坊中心的生态贮流沟
Rainwater ditches at the foot of buildings　06-08　建筑物脚下的生态贮流沟

Rainwater ditches at the foot of buildings | 建筑物脚下的生态贮流沟

061

Atelier de Paysages Bruel-Delmar

Bottière Chênaie Eco-Neighborhood
波提耶尔-申内生态街区

Location：地点
Nantes, France
Completion date：完工日期
2008–2018
Area：面积
30 ha
Client：业主
Nantes Métropole Aménagement
Contracted partner：合作事务所
Confluences, SCE, J.P. Pranlas-Descours
Photo credits：图片版权
Atelier de Paysages Bruel-Delmar

The Bottière Chênaie neighborhood displays its relationship to the territory by its integration into a geological history and its rendition of market gardening. The park strengthens the new link between the neighborhoods, accompanies the new boulevard, following and balancing the new densities of its predominantly indigenous vegetation, and expresses a certain idea of nature. The reopening of the stream enhances the place that finds, in this small geography, a direction and a centrality. The entire neighborhood orients itself towards it. Alleys descend the slopes to the stream, children travel through the fords, passageways span the pontoons where visitors profit from this newly revealed water.

Water and nature, however, are not just found at the park and the stream of the Gohards but are expressed in the dense urban space, and are integrated into the vegetable gardens. This sharing of the public space demonstrates the compatibility between urban density and urban nature. The collection of rainwater coming from the built-up city blocks is done in cultivated ditches. Creating cool areas in the city, this rainwater is essentially expressed through the vegetation of white willows and reeds that occupy half of the new shared thoroughfares.

There are many traces of the systems that have been employed on the site in the past; among them some reservoirs, connected to wells, still remain and contribute to the identity of the place. Besides their conservation, and the restoration of a few of these reservoirs, wind power permits them to renew the supply of water for future shared gardens and the park. This rainwater feeds the wading pond of the central square and, in the southern part of the park, two new ponds transcribe the dictates of the Water Act in an aesthetic space and give the area a high-quality use of water.

波提耶尔-申内生态街区的整治方案被有机地融入了基地的历史与地理环境因素。此街区与新建的林荫道结合，共同建立起了老杜隆街区和波提耶尔街区之间的新连接。城市公园伴随着、也平衡着都市的新密度，其中的植物以当地种类为主，展现出一种在当代久违的自然风光。区内溪流的重新开掘，在这个较小的地域内，增强了邻近地区人们的方向感和中心意识。邻近地区得以以此溪流为基准确定方位。小路沿坡地下降到溪谷中，孩子们在溪流中涉水远足，水滨建立起码头，来此地的参观者从新开发的水流中获益良多。

然而，水这种自然因素并不仅仅在公园和戈阿尔斯溪得到呈现，也在密集的城市空间中被给予了表达，并被整合进花园中。这种公共空间的共享，表明居住密集的城市和自然之间是具有相容性的。城区的雨水收集是通过挖掘出来的沟渠进行的。公路两旁的白柳树和芦苇等植物，受到沟中雨水的滋润，为城区增添了生机。

在这块土地上，我们可以看到众多人类智慧的痕迹。在这其中，几个和水井相连的灌溉用储水池构成了、也强化了这块土地的典型特征。除了保存和重新恢复某些水池的功能状态之外，风力能量也被使用在浇灌用水系统中，有助于水池为未来建成的共享花园（家庭式花园）和公园更新用水。雨水为中心广场的浅水池提供水源，此外，在花园的南边，两个新建的池塘不仅演绎了法国园林将水作为审美空间装饰元素的原则，而且为附近地区提供了高质量的水源。

Nature in the city and urban density at the heart of Nantes：南特中心密集城区的绿化带

Wind turbine and its reservoir above the stream　01　风力发电站及位于溪流上方的附属水库
Crossing the stream, passing the ford　02　横跨溪流、穿越浅滩

The blue fabric accompanies the urban fabric ┊ 与城市肌理结合的水文网络

- Hydrographie principale - Ruisseau des Gohards
- Canal
- Bassins en eau
- Bassins à sec
- Noues végétalisées
- Enrochements drainants
- Parcs / Jardins
- • Puit
- ■ Réservoir
- * Eolienne de puisage

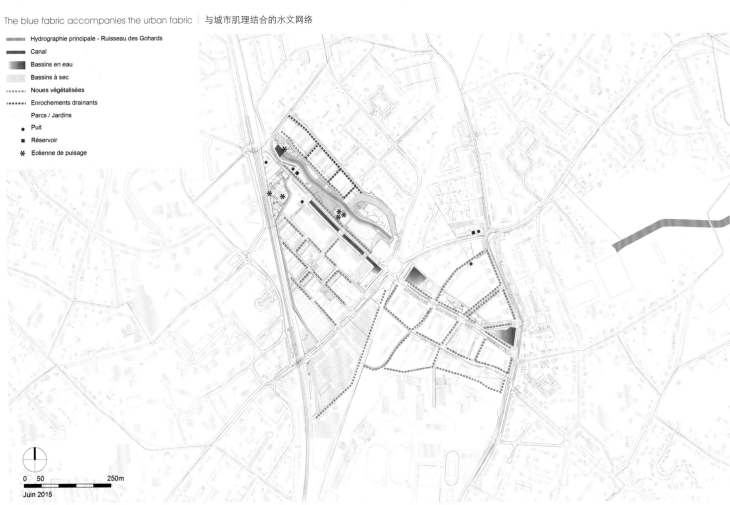

0 50   250m
Juin 2015

Rediscovering the pleasures of water in the stream, uncovered to the open air　03　重新体验在露天溪水中嬉戏所带来的快乐
Living amidst rediscovered water　04　体验重新发现的水世界
Pontoons and gardens in spontaneous vegetation　05　野生植物中间的浮桥和花园
The Gohards Steam, heart of the new neighborhood　06　戈阿尔斯溪成为新开发街区的核心元素

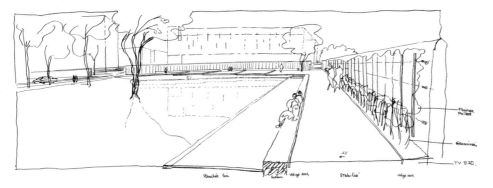

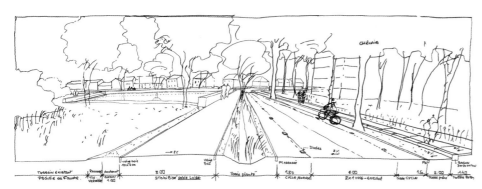

Southern park, two ponds for rainwater collection
南部公园的两个雨水池塘

In front of the school, the canal collects rainwater	07	学校前面收集雨水的运河
The canal planted with vegetation lengthens the park up to the foot of the facilities	08	种植水生植物的运河将公园绿意延伸到公共建筑的脚下
Green and blue fabric all along the promenade	09	蓝绿交织的水渠沿着大道伸展
Iris pseudoachorus, Joncus inflexus, Caltha palustris...	10	运河岸边的黄菖蒲、片髓灯心草、驴蹄草……

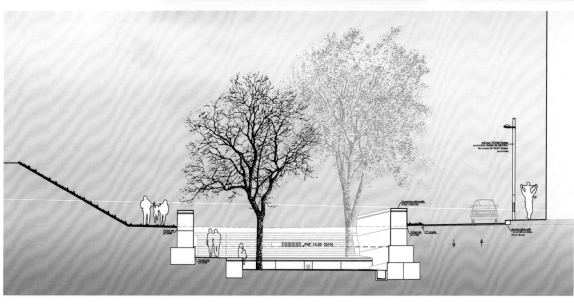

The 'pond of friends':
a designed response

"朋友池塘"：
一个巧于设计的防洪设施

The 'pond of friends' at the walls of the schist gabion	11	以装有片岩的金属笼墙体所围塑出来的"朋友池塘"
The dry pond is an accesible space	12	可以进入的干涸池塘
Expressing water down to the slightest details	13	即使是细节设计也充分表达出水的存在
Outlining water by its structures	14	以建构物来赋予水体形态
Storm spillway	15	暴雨泄洪通道
Oak steps to benefit from the structures	16	橡木台阶可供人们停留休憩，使防洪设施也成为实用空间

Harvesting rainwater creates a new urban landscape | 雨水收集系统造就了新的城市景观

Shared roads 17 有渗排水功能的道路
Rainwater and densities, two objectives reconciled 18 既能排渗雨水又提供了硬质路面，达到两全其美的效果
Here everything expresses the presence of water 19 每种事物都显示出水的存在
A vegetal presence makes water harvesting possible 20 水的收集与植物的生长互利共存

At the foot of the façade, draining ricrac	21	建筑物脚下的排水碎石堆
Playing in the pedestrian paths	22	在步道上玩耍
Planted pedestrian path and slate schist for water collection	23	用板状页岩铺面、具有水收集功能的绿化步道
Coming home from school	24	从学校回家
Little pedestrian path: water collection and density	25	提供硬质路面并有集水功能的小行人步道

Atelier de Paysages Bruel-Delmar

La Morinais Neighborhood
莫利奈街区

Location | 地点
Saint-Jacques-de-la-Lande, France
Completion date | 完工日期
2015
Area | 面积
40 ha
Client | 业主
Ville de Saint-Jacques-de-la-Lande
Contracted partner | 合作事务所
Cabinet Bourgois, J.P. Pranlas-Descours
Photo credits | 图片版权
Atelier de Paysages Bruel-Delmar

The Morinais neighborhood is part of a comprehensive urban project that is rooted in a particular territory. This dense neighborhood and the ecological park, which is its counterpoint, are part of a vast experimental laboratory for the collection and treatment of rainwater. Started in the 1990s, the project has made it possible to elaborate the principles and the precursor vocabulary of the 'zero pipes' method for the management of runoff water. The method of draining a street, a garden, or a parking lot affirms the geographic position of a place. When water is considered a resource of the urban project, high-quality development of public and collective spaces is made possible by using the money that would otherwise have been spent for pipes. The argument is simple and efficient, and has convinced the elected officials who brought forward this challenge.

This shared attitude has made it possible to shape the design of the park early on in the process via the creation of a settling tank and the general design of a bridge, which reveals the valley from the tiny Blosne stream. Runoff water merges with the course of the stream, enhances the hillside, and creates a link between the dense urban center and the ecological park that surrounds it. The ponds of the old farms are preserved and become new centers of diversity and a way of maintaining nature in the city. Rather than a stereotypical response, this water collection system is custom-designed, making it possible to set forth a vocabulary of small masonry, taps for the disposal of water, aqueduct thresholds, and head works in prefabricated concrete, red schist forming spaces that drain into the banks of shared pathways and narrow channels. This work reveals the presence of water as the tool of an urban project situated in, and conveying the expression of, its territory…

To permit this subtle, gravity-based management of water flow, the contracting authority proposed to private operators that the leveling and development of the centers of private blocks should be designed by public-space landscapers. This local-to-global management has helped control the path of the water from the gutter to the river.

莫利奈街区设计是一个综合性城区设计方案的一部分,这一设计是建立在所在区域的独特性基础上的,当地的发展受惠于其地理优越性。这个人口密集的街区和邻近的生态公园相映成趣,成为一个具有代表性的收集和处理雨水的大型实验室。项目启动于20世纪90年代,先期进行了精心的规划,以"零管路"口号作为施工先决条件。街道、花园或停车场内的排洪被设计为利用地理高度自然疏导的方式。水在这一城区设计项目中被看成一种开发高品质公共和集体空间的资源,在前述地表环境中的投资绝不应用于购买和安装管道。这种观点是简单而有说服力的,已经影响了需要具体实施项目的当地民选政府。

这一主流态度导致了一系列早期设计方案的成形:建造一个沉淀池,并在小波尔斯尼溪上方建造一座桥,以便俯瞰下面的山谷。雨水裹挟着溪水下流,使得溪岸变得更加坚实,在人口密集的城市和围绕着城市的生态公园之间建立起联系。旧农场中的池塘被保留下来,作为保持城市自然风格的手段,成为园区内体现多样性生态的中心。相对常规的设计方案,本项目水收集系统的定制设计属性更强,使得设计有可能被概括为一系列关键词:小型砌体,给排水水龙头,预制混凝土构件的水渠入口和主要设施,将雨水导入道路侧面的红色片岩空间,狭窄通道。这项设计揭示了水作为城市设计项目依存的环境因素所具有的表达能力和疆域范围……

为了达到借助重力进行水管理的效果,委托方提出私营运营商应负责平整土地,并且由承担公共空间景观设计的事务所对街区内的集中私人产业开发进行统一设计。这一水管理模式具有从局部地区推广到全球的价值,使人们对于从排水沟到河流的水体的控制成为可能。

Facing the new city centre, the rivulet leads the water towards the river

朝向新城市中心、将雨水导入河流的小溪

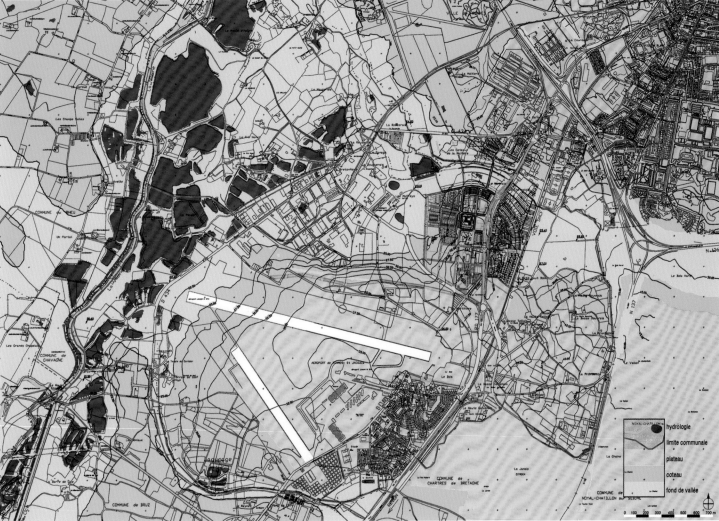

The geomorphology directs the development of the territory ┊ 地形影响区域的开发

The ecological park is a counterpoint to   01   生态公园和空间密集的
the dense city centre   城区形成互补

075

The 40 hectares of park constitutes the first facilities of the town
占地 40 公顷的公园成为新建城市的首要公共设施

A hydrography at ground level characterises the ecological park　02　地面上的浅水滩为生态公园增添了特色
The basin lined with oaks, artificial structure for sedimentation　03　岸边种植成行橡树的人造沉降水池

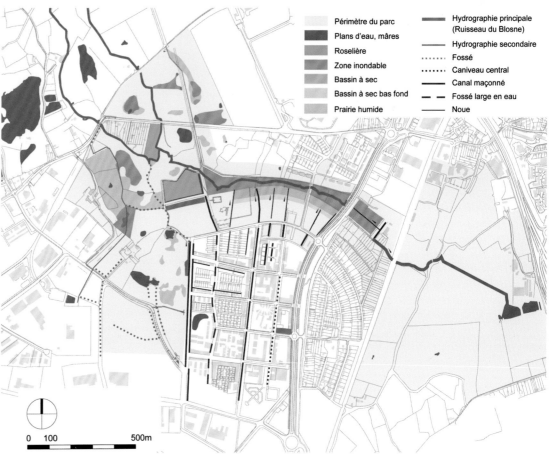

Périmètre du parc	Hydrographie principale (Ruisseau du Blosne)
Plans d'eau, mâres	Hydrographie secondaire
Roselière	Fossé
Zone inondable	Caniveau central
Bassin à sec	Canal maçonné
Bassin à sec bas fond	Fossé large en eau
Prairie humide	Noue

0 100 500m

A preserved and amplified blue framework; a route from the gutter to the river
基地中被保留和强化的水文结构；从水渠到河流的历程

The artificial reedbed and its observation pontoon	04	人工芦苇地和观测浮桥
Pontoon on the 'iris pond'	05	鸢尾属植物池塘上的浮桥
The reedbed secures the treatment of runoff water	06	芦苇地确保了地表径流水的净化处理
The willow garden, a salon in the middle of nature	07	柳树花园，位于大自然当中的休憩厅堂
Promenade at the border of the reed bed	08	芦苇地边上的散步道
High meadow and pasture	09	高草地和牧场

In the parking lot, tricycle around the canal of rainwater collection 10 停车场内儿童绕着雨水收集渠道骑三轮车
Between park and the city centre, a road flanked 11 连接公园和城市中心的道路，
by a rainwater collection ditch 其侧面设有雨水收集沟
The commercial centre's parking lot, planted with vegetation 12 绿化的商业中心停车场

At the foot of the roadway-dike and the centuries-old oaks, a freshness supporting biodiversity	13	在堤道和百年橡树脚下，阴凉的环境有利于生物多样性的发展
Structures of red schist and concrete to highlight the path of the water	14	红色页岩和水泥搭配建成的各种结构体点缀着水流的路径
Crossing and stairs towards the roadway dike	15	穿越和通往堤道的阶梯
The big ditch of the roadway dike	16	堤道旁的大沟

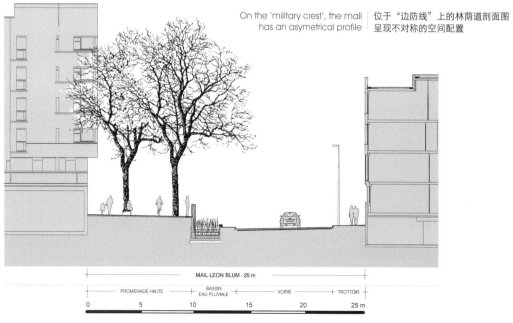

On the 'military crest', the mall has an asymmetrical profile | 位于"边防线"上的林荫道剖面图呈现不对称的空间配置

Léon Blum Promenade and its rainwater collection canal　17　雷昂·布朗林荫道和收集雨水的沟渠
A dry canal planed with phragmites　18　种植芦苇的干沟渠
Promenade under the oaks of the pedestrian thoroughfare　19　在人行步道的橡树下散步

18

Permeable soil and seed traps in the crushed paving	20	被翻覆捣碎的沥青路成为可渗透地面和可让植物种子附着的环境
'Re-naturing' the former departmental road and using it for soft transportation, cycles and pedestrians	21	一条重新恢复自然环境的旧省级公路，成为"慢行交通"（自行车、步行、滑轮等）的使用空间
Block of freshness in the earth	22	绿意盎然、阴凉清爽的街坊中心
Onto the garden, gravity-fed irrigation, perched on the roof of the commercial center	23	购物中心屋顶花园的重力灌溉设施
Collection of runoff water by a formation of the soil	24	借助地面起伏形态来收集地表径流水
Reading the path of the water: Dauphin precast concrete	25	从预制水泥排水口的设计来阅读水流路径

Simple lawned ditch under the wild-cherry orchard 26 野樱桃园林下铺植草地的简约沟渠
Water at the surface of the soil promotes biodiversity 27 土地表面的浅水滩促进了生物多样性
Crossing the ditch over a ford to reach the centre of the block 28 穿越浅滩中的水沟而抵达街坊的中心

Making the water course a shared space	29	借助水流的经过创造共享空间
Collecting water in the middle of dense city blocks	30	在高密度的城市街坊中收集水体
A permeable landscape that follows the path of the water	31	与水道相伴的绿化景观
At the heart of the city blocks, the vegetation orientates the eye towards the park	32	种植于城市街坊中的植物将人们的目光导向公园
Permeable soil permitting a fertile freshness	33	渗水地面创造出清爽肥沃的生物环境

Coulon Leblanc & Associés

Croix Bonnet Neighborhood
克鲁瓦·伯奈特街区

Location | 地点
Bois d'Arcy, France
Completion date | 完工日期
2013
Area | 面积
35 ha
Client | 业主
Agence Foncière et Technique de la Région Parisienne
Contracted partner | 合作事务所
Hydratec, 8'18''
Photo credits | 图片版权
Coulon Leblanc & Associés

Rather than seeing it as an unfortunate problem, it is always interesting, even essential, to treat the question of water as the dynamic of a project, whether it is urban or for an entire territory. With a focus centered around the management of water, the project will become better, more geographically situated, and more powerful. This indispensable program imposes solutions that create the opportunity to de-trivialize conventional landscaping interventions. Thinking from this technical point of view, the opportunity arises for a site-rooted approach and, as a consequence, its aesthetics avoid the notion of structural embellishment, which is a good thing. Decor today is certainly one of the principal enemies of the landscape.

The project of the la Croix Bonnet neighborhood is situated on a historical waterline. The channel of the Clayes is a work that was created on the orders of Louis XIV for feeding the pools and fountains of Versailles. The question of water was first and foremost a technical necessity of site preparation. A canal was dug to manage the

当项目中牵涉到可以将其作为一个动态变量的水的问题时，项目就变得与其说是麻烦，还不如说总是非常有趣，甚至是项目不可缺少的元素，无论这个项目是在城市里还是其他地域空间。围绕水的管理对项目进行深思熟虑，会让该项目变得更完善，更加适合其所在的地理环境，更加严密完备。在实施这个必不可少的程序后提出的解决方案，往往有更多机会去摆脱传统园林设计思想的窠臼。从技术角度来说，这种思考方法给人一种植根于项目基址思考的机会，从而避免在美学上做出概念化的结构修饰设计，这是一件好事。然而在今天，装饰是景观设计的主要敌人之一。

克鲁瓦·伯奈特附近的这个项目位于一片历史渊源很深的水域旁边。克拉耶运河最初是按照法王路易十四的命令建造出来，为凡尔赛城堡的喷泉和水池提供水源的。水是城堡中的喷泉和水池能够得以运作的技术性必要条件。这就

On the traces of the Clayes rivulet dug out to feed the water works of the park of Versailles, the canal is a great classic
运河位于古代挖掘出来为凡尔赛花园提供水源的克拉耶水渠基址上，堪称伟大的经典之作

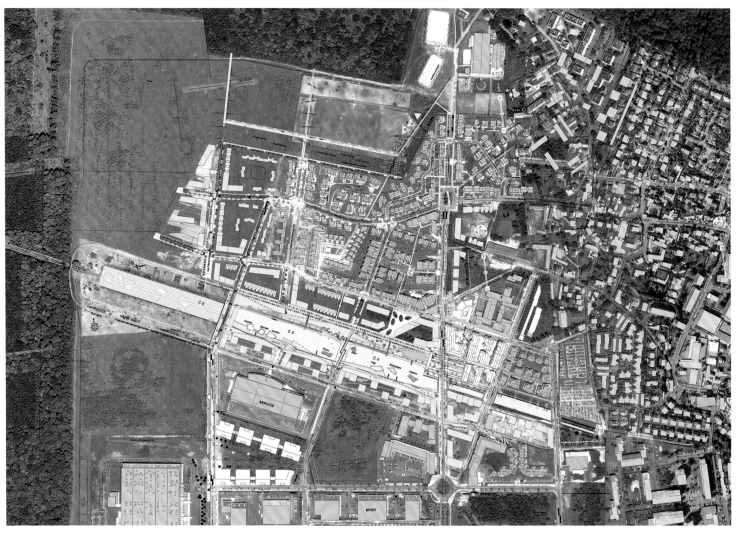

rainwater of a 165-hectare site that made a new neighborhood possible, with the Fonciere and Technical Agency of the Parisian Region the developer. The rainwater has to be contained in ponds and released slowly in order not to saturate the outlets. It also afforded the possibility of staging the whole site around a series of six large ponds, allowing a spectacular, identity-creating axis that can be exploited by the promoters to add value to the housing.

Upstream, this created a new ecological wetland situation with the gradients of humidity linked to the slopes of the banks promoting biodiversity. Seven hectares are listed in the Natural Zone of Ecological, Faunistic, and Floristic Interest. This space, a unique location in the middle of a neighborhood, is enclosed and everything is organized around it to showcase it and render it readable. A long balcony walk and a 12-meter-high vista point provide information on the fauna and flora. Further downstream, the pools are accessible, flanked, or traversed by wooden piers, pontoons, and bankside promenades, which make this urban space seem a real park.

The variation of the water level of the pond makes the visitors feel closer or further away from the reflecting pool. Heavy showers threaten the continuity of the passageways or the accessibility to the works. Water permeates the entire site and reflects the light of the sky; the rest supplements the formal agenda to develop the course, places, and the environments.

要求挖掘一条运河，以汇聚邻近165公顷地区的降水，这一工程使得邻近地区的面貌为之一新。巴黎地区土地和技术局成为该项目的开发商。雨水将被蓄积在池塘中，缓慢地被释放出去。这一个工程使得当地的六个大池塘得到了集中的展示，创建者随之将为各个池塘编写确认其身份的伟大家史，以便为设施增加价值。

所有这一切的努力在运河上游创造出来一个新的湿地生态环境，湿地的湿度与运河的斜坡一起，促进了该处生物多样性的发展。这就是面积为7公顷的法国生态、动物和植物自然保护区。这个区域的情况在附近地区中间是罕见的，空间是封闭的，区域中的一切都是围绕水来展示的。区域内设有一个长阳台式步道和一个12米高的远景瞭望台，可以观测区域内的动物群和植物群的情况。运河下游有很多池塘，岸边或水面上有木栅、浮桥和河岸长廊，使得这个空间具有浓郁的城市格调，不愧是一个真正的公园。

池塘水位的变化使得参观者更为接近或远离水面，暴雨偶尔威胁着通道或阻碍人们进入项目所在区域。水涵养着整个区域，映照出天空以及其他一切，除了所行经的旅程、地方和环境在心中留下的点点印迹。

This ZNIEFF (Natural Zone of Faunistic and Floristic Ecological Interest) is the happy consequence of the necessity to regulate the site's water. A strong potential in terms of biodiversity with a rigorous geometry: two characteristics that are not incompatible

01 该动植物生态保护区的建立系出于调节项目所在区域水资源的目的，其建成是令人高兴的；
在严格的几何构图下，生物多样性的发展仍然十分具有潜力，
显现出这两者之间是可以互相兼容的

Multiple uses: areas for walking, for contemplative calm 02 空间的多种用途：既可供人散步，亦可作为宁静思索的场所
The waterworks multiply the look-out points over the water: this wide 03 供水设施为人们提供了众多的水面观察点，
canal that is an ornamental pond is however above all, a system 宽广的运河在起到雨水储存作用之前，
for rainwater retention 首先也是一种景观

04 In the foreground, a very urban promenade on the water's edge; in the background, a passageway that links the northern and southern parts of the neighborhood. The weathervanes speak to us about wind

05 Wet zone plantings, along with educational information, isolate the promenade from the bank of the canal

04 眼前是一条设置在水边的城市散步道,背景中有一条连接南北街区的桥道,风向标则向人们诉说着风的故事

05 深入湿地生态环境的散步道上设有教学信息,其周围种植的植物将散步道与运河岸隔离了起来

Atelier Villes & Paysages

Champ-Fleuri Neighborhood
查姆普-弗瑞街区

Location：地点
Bourgoin-Jallieu, France
Completion date：完工日期
2013–2016
Area：面积
5.5 ha
Client：业主
Communauté d'agglomération des portes de l'Isère
Contracted partner：合作事务所
Egis
Photo credits：图片版权
Atelier Villes & Paysages /
Renaud Ducher & Laurie Richard

Requalification of the public spaces of the Champ-Fleuri ZAC (comprehensive development zone) is part of an operation led by the National Agency for Urban Renewal. It is, above all, an operation of urban renewal of a neighborhood of low-cost housing of which the principal themes are: attractiveness (for residents, businesses, and public facilities); integration (of the neighborhood in relation to its urban and landscaping context); connection and weaving (neighborhood with city, spaces with users, river dwellers with one another, economic activities with housing); and development (public spaces, roadways, parcels, and private spaces).

All of this offers to the sector an enhanced image, including a strong notion of the appropriation of space, an image with which the users and residents are now proud to identify. Around a central park in the process of being created, a green framework is extended into the redeveloped streets in the form of a linear garden that borders soft means of transport. The planted and mineral spaces mingle due to paving stones separated by plantable joints that directly infiltrate rainwater.

The surfaces of pedestrian thoroughfares include Kronimus concrete paving stones. These paving stones go all the way to the center of the planted strip to provide infiltration. Separating frames built into the paving stones create a joint of four centimeters that is filled with earth and seeded. Under the paving stone, a foundational structure has been put in place that is both load-bearing and draining, sized to accommodate a volume of water storage. The paving stones with these small joints planted with vegetation are installed at the lowest point of the promenade, collecting and infiltrating the runoff water and eliminating the need for an underground storm drainage system.

查姆普-弗瑞综合发展区公共空间的改造项目,属于法国城市改造署(National Agency for Urban Renewal)组织改造项目中的一部分。该项目首先属于廉租住房社区的改造,主要侧重方面是:对居民、企业和公共设施建设的吸引,邻近地区城市环境和景观环境的整合,整合附近地区和城市、居住空间与居民、河畔居民之间的联系和组织,以及公共空间、道路、货运和私人产业的发展。

这些规划一旦实现,将极大改善所述扇形区域的空间利用状况,增强当地居民和设施使用者对社区的认同度。在改造区中已建的中央公园附近,一片绿色带以条状花园的形式,点缀着道路两边。这些植物的栽植和空间利用的经济性,得益于道路的铺筑使用了特殊的砖石,在道路中间保留了可供种植、能够直接渗透雨水的间隙。

人行道路面铺筑所使用的材料是克罗尼马斯(Kronimus)水泥铺路砖。所有的绿色植物带路面使用的都是这种石材。铺路砖中间有四厘米宽的空隙,以便容纳土壤,进行播种。在路砖下面设置的基础设施能承受相当程度的负载,并起到排水作用,负载大小和排水能力多少因地下储存水的多少而不同。铺设这种路面的时间段选择在地面雨水流动、积蓄和渗透量最小的时候,以避免施工受到地下水的影响。

Map of the redeveloped neighborhood: new central park, streets transformed into garden promenades

重新开发的街区：
新建的中央公园、街道被改造成花园散步道

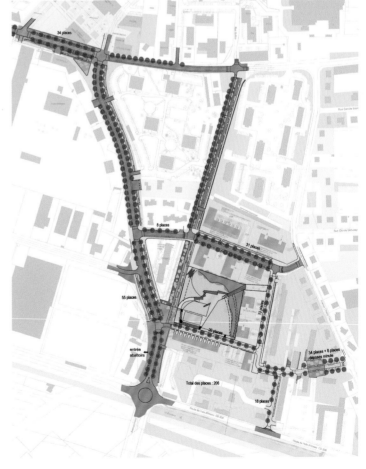

The road, the bike path and the pedestrian promenade are separated by permeable bands of flowered meadows and linear gardens
Situated at the lowest point, the strips of linear garden are infiltrating

01 车道、自行车道与人行步道之间
以花卉草地和长条状花园作为隔离
02 位于较低地势的长条状花园具有渗透水分的功能

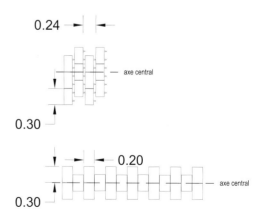

Detail of the lay-out of the paving stones with and without plantable joints　带有种植缝隙和不带有种植缝隙的铺石地面细部大样

According to the seasons and the intensity of the trampling it sustains, the vegetation of the joints sometimes cover a large part of the paving stones　03　铺石缝隙中的植物生长情况依据季节和踩踏程度而不同，有时可以很茂盛地覆盖大面积的铺地

The infiltrating paving stones aid the circulation of the maintenance vehicles　04　具有渗水性的铺石路面上可以行驶清洁维护道路的车辆

Implementation of the paving stones with plantable joints　05　带有种植缝隙的铺石地面，施工中状态

The project recalibrated the roadways, pushing the road far from the residences so they could be bordered with garden promenades 06 方案重新调整了道路位置和尺度，使其与住宅建筑保持较大距离，并因此可在路边设置花园步道

The strips of permeable gardens include furniture useful for people taking a break 07 条状的渗水花园里面设置了座椅可供人们小憩

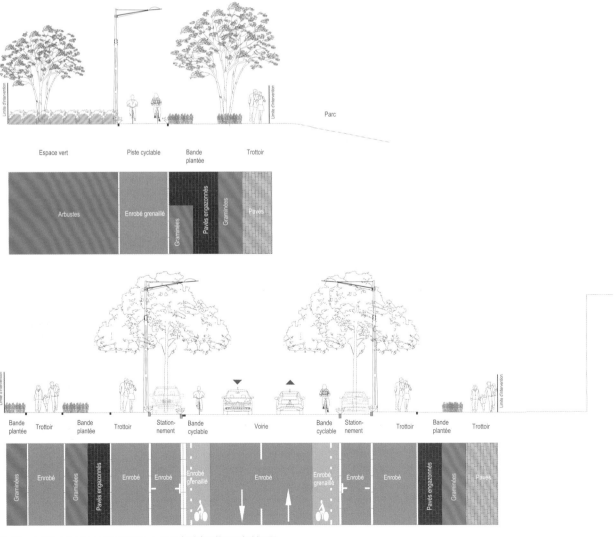

Profiles of different types of roadways ┊ 不同类型路面的平面与剖面图

| Technical profile on the structures of the roads and the permeable areas | 道路结构和渗水区的技术剖面图 |

The infiltrating paving stones are interrupted irregularly to accommodate trees and shrubs 08 规律性地嵌入乔木和灌木丛的渗水铺石路面

The hardy vegetation is adapted to the alternation of dry and wet periods 09 适应旱季和雨季转换的耐旱植物

The principal axis, Scotto Boulevard, is bordered with strips of flowered meadows 10 作为项目主要轴线的斯科特大道，周边是条状的花园草地

Atelier Ruelle

EuroNantes Gare/Malakoff Neighborhood
欧洲南特火车站 / 马拉科夫街区

Location 地点
Nantes, France
Completion date 完工日期
2004–2018
Area 面积
164 ha
Client 业主
Nantes Métropole, Nantes Métropole Aménagement
Contracted partner 合作事务所
Océanis
Photo credits 图片版权
Atelier Ruelle

The development of public spaces is at the heart of the transformation of 164 hectares of the EuroNantes Gare/Malakoff neighborhood. At the beginning of the project, in 2002, the Ruelle Workshop proposed to the city of Nantes the idea of combining the development potential linked to the new TGV high-speed railway station with the opening up of the Malakoff neighborhood. This large, low-cost housing structure on the edge of the river was isolated by the railway infrastructures and industrial wastelands, making it difficult to get to and from the center of town.

The presence of natural elements, and more particularly water, orients the project. The future neighborhood opens up towards the banks of the Loire to the south, towards the Saint Félix Canal to the west or towards the Little Amazon Rainforest to the east (Natura 2000 wetlands). Situated in the flood plain of the Loire, this neighborhood is built historically on soils backfilled with sand coming from the river (3 meters of backfill). Sand is the material of choice because of its infiltration capacity, allowing the water to diffuse into the soil. The project capitalizes on this characteristic for developing a particular way of managing rainwater.

公共空间的整治是占地164公顷的欧洲南特火车站/马拉科夫街区的核心项目。项目的开始是在2002年，街巷工作室（Atelier Ruelle）建议南特市将潜力巨大的TGV高速火车站发展跟开发车站附近的马拉科夫街区相结合。当时，这片位于河边的庞大社会住宅区被周围的铁路基础设施和工业废墟所孤立，因此与市中心的交通非常不便。

自然因素的存在，特别是水的存在，极大地影响了项目的规划。未来的社区南朝卢瓦尔河，西向圣菲利克斯运河，东对小亚马逊雨林。社区坐落在卢瓦尔河的冲积平原上，所在区域的土壤来自河沙回填（回填3米深）。沙子是优良的渗水材料，沙子的渗水过滤能力允许水扩散到地面。该项目利用当地的这一特点，发展出一种特别的管理雨水方式。

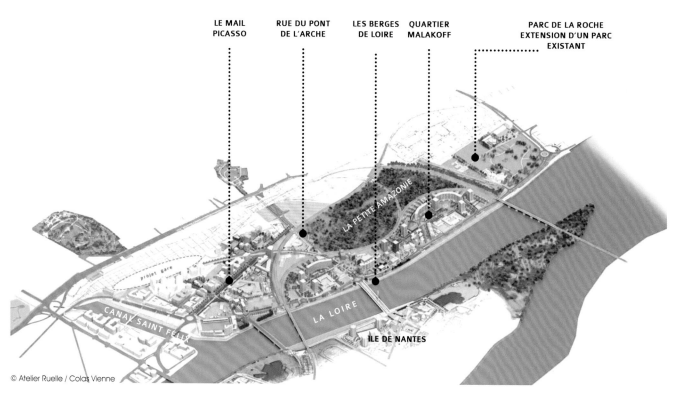

EuroNantes Train Station and Malakoff: axonometry of the state of progression | 欧洲南特火车站 / 马拉科夫街区空间分布透视图

The collected rainwater is sent through perforated pipes that allow the water to spill into the sandy soils. With this principle of purification, the volume of water to be stored in the retention basins is reduced. To reinforce the permeability of the soils, the project also develops the utilization of porous materials: grass-covered paving stones and permeable road surfaces.

收集的雨水通过穿孔的管道传送，这种管道可以让水中沉淀下来的沙子重新回到到沙土中。由于水得到了净化，进入到滞留池中储存的水量减少了。为了加强土壤的渗透性，该项目还开发出了多孔材料的利用方法：草地覆盖的铺路石路面和透水路面。

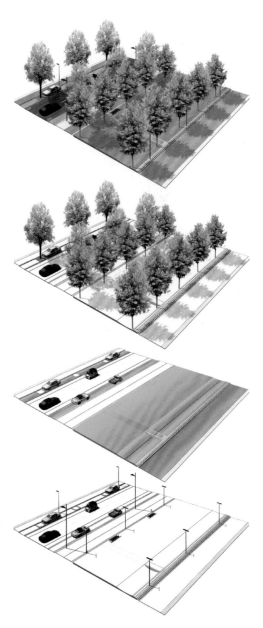

Picasso parkway: the structuring axis of the EuroNantes Train Station neighborhood is a large infiltration trench in sandy soil for all the pedestrian public spaces

毕加索公园大道：此乃欧洲南特火车站街区的结构性轴线，大道上的所有步行公共空间皆为沙土地面，形成一条大型渗水堑道

01 Picasso parkway: above, playgrounds, school parvis, below, a draining trench

毕加索公园大道：地面上设置了儿童游戏场、学校前庭广场等空间，地面下则是排水堑道

From the neighborhood, rediscovering the banks of the Loire　02　从街区中眺望整治后的卢瓦尔河岸
Integration of the banks in the larger landscape of the Loire　03　将河岸整治融入卢瓦尔河的整体景观中
Before: the banks of the Loire show the presence of the roadways　04　项目整治前：被道路占据的卢瓦尔河岸
After moving the roadway, the permeable surfacing characterize　05　将道路移开后的卢瓦尔河岸
the new promenade along the banks of the Loire　　　成为具有渗水性的散步道

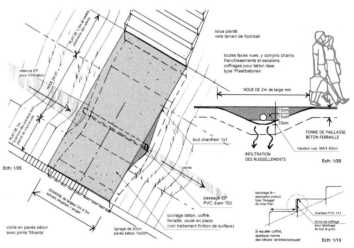

Detail of the technical aspect of the ditch and its crossing near Mauves Park | 技术细部图：位于莫沃公园附近的生态贮流沟和沟上的跨道

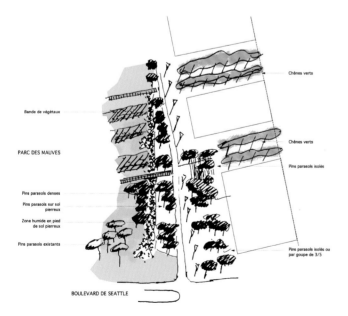

A play of crossing permits the fording of the water zones | 一系列跨越水渠的设施，带人穿越潮湿地带

A draining ditch of the large roadway delineates the boundary with Mauves Park 06 大路旁的生态贮流沟，勾勒出莫沃公园的边界
The principal access road to the La Roche Park is flanked by a ditch 07 通往拉罗什公园的主要通路边设置了生态贮流沟
Malakoff Neighborhood: the water course of the old street is preserved, 08 马拉科夫街区：旧街道的排水渠被保留下来继续利用，而停车场的不渗水地面
the impermeable soils of the parking lots are reduced to the actual spots for the cars 也被减至最低量，仅限于停车位的范围
At the centre of the Grand Ensemble, more trees, more landscape 09 在大型社会住宅区的中心，种植了更多的树木，创造了更多景观

When the neighborhood encounters the wet zone of la Petite Amazonie (Natura 2000), the boundary blends into the environment very naturally
A promenade that looks out over the Petite Amazonie makes it possible to contemplate this wet zone

10-11 新街区与小亚马逊河湿地的交接地带的处理
12 小亚马逊河上的散步道观景台,让人得以观赏此一湿地生态环境

Atelier Ruelle

Baudens Neighborhood
鲍登斯街区

Location：地点
Bourges, France
Completion date：完工日期
2009–2019
Area：面积
5 ha
Client：业主
SEM Territoria
Contracted partner：合作事务所
BIO TOP, SAFEGE
Photo credits：图片版权
Atelier Ruelle

Covering an area of about five hectares, this neighborhood transformation is situated near the town center of Bourges, along the 19th century boulevards. A former military hospital, the Baudens neighborhood takes its inspiration from its past. From an enclosed area, removed from the public fabric of the city, it becomes a neighborhood in itself. The heritage buildings are renovated (under the control of the ABF, an organization of French architects) and are 'recycled' for a new use: the city is constructed on the city.

A limestone site filled with crevices, the soil is conducive to infiltration. The project implements a system of transportation ditches in order to limit the development of underground networks. Each pedestrian path or roadway is therefore systematically accompanied by a planted ditch that collects its runoff water.

The impermeable parts of the project are limited to the strictly necessary (a parking lot and the very reduced principal service road) in such a way as to promote direct infiltration of rainwater into the water table. The porous soils, helpful in the preservation of the existing trees, accommodate leisure areas: the Garden in the Wind and the Court of Honor. The latter, made of an undulation of earth 40 centimeters high, which is the height of a seat, makes it possible to store a part of the rainwater. The entire system of alternative management of rainwater permits the reduction of overflow drainage to existing networks by 50 percent.

该街区约有 5 公顷的面积，位于布尔日市中心地带附近的 19 世纪大道沿线。作为从前的军事医院，鲍登斯街区的改造重新利用了许多历史性元素。这块原本不受城市肌理所影响的封闭军事用地，逐渐转变为融入城市的街区。而基地中的历史性建筑，则在拥有 ABF 头衔的法国建物管理建筑师的管控下，被重新整治、赋予其它用途，此乃所谓"在既有城市上继续建造城市"。

基地的石灰质地面充满裂缝，非常有利于水的渗透。为了限制过多的地下工程，该项目设计了一个地面上的生态沟渠系统，用于水的管理。每一条人行道都系统地伴随着一条种有植物的沟渠，以便收集地面的雨水。

不能渗水的地块被严格限制范围（停车场和缩减到非常短的服务性主路），以便使雨水尽可能地渗入地下水系统中。多孔的土壤有助于保护现有的树木，为人们提供令人愉悦的空间——"微风花园"和"荣誉庭"广场。"荣誉庭"广场具有 40 公分的起伏地形，也就是一个座位的高度，这使得这种花园可以存储一部分雨水。此项目所设置的所有替代式雨水处理措施，使原本必须通过现有地下管道系统来进行的排水，可以足足降低 50% 的水量。

| LA PLACE BAUDENS | LE «JARDIN SOUS LE VENT» | LA COUR D'HONNEUR | The regular framework of the former military hospital becomes the landscaping 'bones' of the new residential neighborhood | 昔日的军事医院规律的空间肌理变成了今天住宅小区景观的主要结构 |

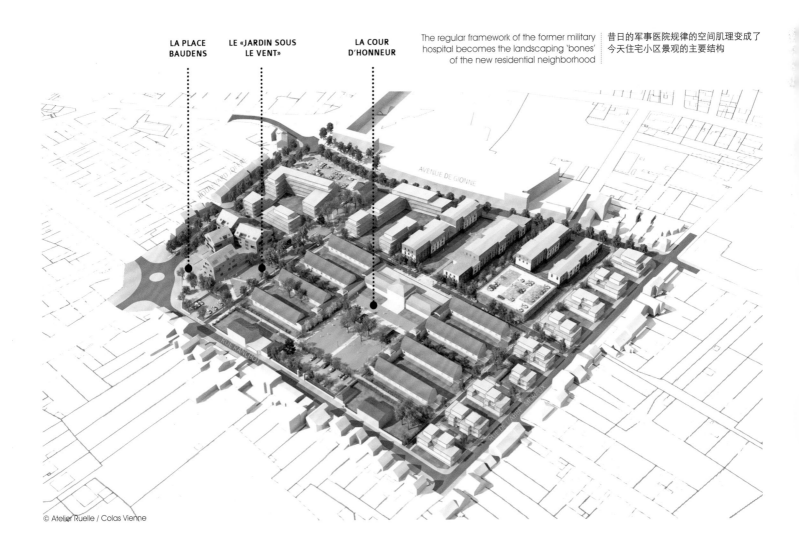

© Atelier Ruelle / Colas Vienne

01 A porous soil, helpful to the preservation of centuries-old trees
02 The court of honour, preserved in its dimensions, becomes a space for relaxation in the heart of the neighborhood

01 多孔路面有利于有着几个世纪树龄古树的保护
02 完整保留下来的"荣誉庭"广场成为街区的休闲活动中心

The court of honour, originally dominated by stonework, rediscovers a porous soil 03-04 昔日硬质地面为主的"荣誉庭"广场被改建为多孔渗水性路面
The central roadway is preserved, from now on with a porous soil 05 被保留下来的中央道路,从今以后也是多孔路面
From the most impermeable to the most permeable, variation according to the uses 06 根据空间用途的不同来处理地面的渗水性
Infiltrating rainwater, filtering the views between the court of honour and an aerial parking lot 07 位于"荣誉庭"广场和露天停车场之间的景观空间,既可渗透地面雨水,又可过滤人们的视线

Atelier Ruelle

Gaubert Building Complex
高贝尔综合小区

Location：地点
Angers, France
Completion date：完工日期
2006
Area：面积
1.35 ha
Client：业主
Angers Habitat
Contracted partner：合作事务所
ACI
Photo credits：图片版权
Atelier Ruelle

A soft residential development project that addresses the landscape and the recovery of the ground floor, and gives attention to the limits and continuities that it shares with the city, the Gaubert Building Complex is composed of 260 apartments distributed into four housing towers. This project is developed around the gardens, the paths, and the parking area, responding to the exigency of quality of use and of landscaping. A new building for the agency of the HLM (French government-subsidized housing) is integrated into the complex.

The landscaping development is based on the path of the water, the diversity of vegetal environments, and the use of schist that resembles the walls of nearby neighborhoods. The footpaths, made of strips of schist placed in trenches, preserve the permeability of the soil while conserving a mineral character, providing the users with both ease and charm.

The ditches—ponds with soft slopes planted with trees—serve a two-fold purpose. Offering an interesting landscaped environment, they also contribute to the management of rainwater, while making it possible to separate the ground-level apartments to give them some privacy.

在住宅开发过程中强调和谐的项目要求设计面向景观和土地的治理，关注建筑物的使用极限和连续性，以便让建筑物与城市的存在周期同步进退。高贝尔综合小区分为四座塔式高楼，有260个单元。大厦周边设有花园、道路和停车场，以跟周围景观达到充分的协调。小区内房屋一部分是法国政府给予补助的经适房。

园林绿化设计是基于水的流经路线，园内种植了多样的植物种类，一些设施建设则采用了跟附近地区墙壁类似的页岩。人行道路面上铺着页岩条石，不仅保持了土壤的通透性，而且使得矿物的独特性得到展示，为居民提供了交通便利和审美空间。

沟渠、池塘的缓坡上栽有树木，可以达到双重目的。首先是提供有趣的景观环境；其次也有助于雨水的管理；此外，这些树木也可以为地面层公寓内的住户提供遮蔽，以保护其隐私。

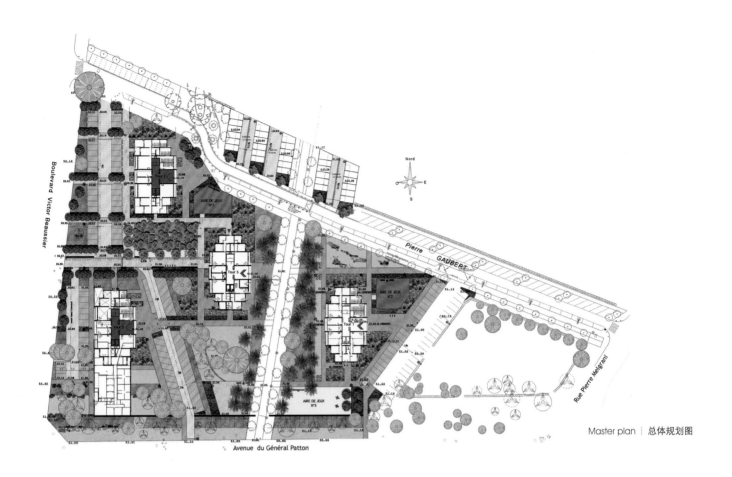

Master plan 总体规划图

The dry ponds in soft slopes put the roadway at a distance 01 具有缓坡的干式浅沟使得道路能够
from the residential towers and the ground floor apartments 跟高层住宅楼和位于地面层的公寓保持较大距离
Rainwater management serves 'residentialization' 02 服务于住宅区改造的雨水管理设施

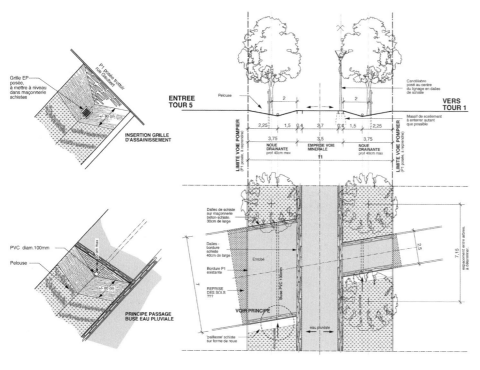

Detail of the layout of the ditches : 生态沟渠细部大样

At the crossing with the transversal pedestrian paths, 03　生态沟渠与人行步道相交处，
the ditch receives a surfacing in schist 　　采用页岩作为沟上跨道的铺面

The strips in schist, placed with wide joints, creates a crossway meant for the residents 04 居民惯用来穿越绿地的小径，被改造为具有大缝隙的页岩条石路

For an economy of management, the gardeners recuperate the runoff water of the roads 05–06 为了花园维护管理上的便利，园丁直接收集雨水径流来进行灌溉

From the large avenue: passageway towards the residence in a play of shadow and light 07 从大道通向住宅区的小径笼罩在一片光影交错的氛围中

The Management of Rainwater in City

城市雨水资源管理

Atelier de Paysages Bruel-Delmar

Haute Deûle River Banks
上德勒河岸

Location : 地点
Lille – Lomme, France
Completion date : 完工日期
2015
Area : 面积
25 ha
Client : 业主
Soreli
Contracted partner : 合作事务所
Profil Ingénierie, G. Pilet, Agh, Venna, Stucky
Photo credits : 图片版权
Atelier de Paysages Bruel-Delmar (n°01, 04, 07–18),
Yves Bercez (n°02–03, 05–06)

The development of the new neighborhood of the Haute Deûle River Banks relies on an appreciation of the qualities of the site, forming the basis of the project and of its enrichment. The presence of water is undeniable, both in the history of this quarter and in its current configuration, despite its loss of status. The project for the Haute Deûle River Banks is based on traces of this memory in order to apply itself to continuity in the identity of inhabited places, and at the same time to revitalize this expression of unifying water in the new development.

Across the large neighborhood the water is omnipresent, finding a place in the public spaces and directing the urban composition. The square around the water station offers a convivial space in the vestiges of the river transport industry that forged the identity of the Bois Blancs neighborhood. Around Euratechnology, the layout of the streets and walkways is dictated by the major north-south orientation towards the Bras de Canteleu Canal and affirms the new urban fabric. The rectilinear paths—planted with ash trees, the emblematic tree of the old marsh—accompany canals and impermeable ditches. The hydraulic devices are completed according to the orientation of the Deûle Valley by the creation of sunken gardens, which harvest the runoff water, and are planted with hygrophyte vegetation that creates natural continuities according to a concept of 'garden-street'.

上德勒河岸协议开发区的整治计划强调对基地环境品质的重新认识，以这些环境特质来建立起方案的坚实基础和丰富性。无论是就历史或现状而言，水在这个基地上存在的重要性是不可否认的，尽管人们逐渐失去了对它的认识。上德勒河岸的规划方案就是循着这个有关于水的记忆与痕迹而发展，一方面借此延伸旧使用场所的识别性，一方面在新的整治中更新水体的表现形式。

在广大的邻近地区，无所不在的水最终汇聚进公共空间内的一个去处，并以此影响了附近社区的结构。水运站前的广场为人们提供了一个交流场所，带有往昔河流运输业留下的痕迹，正是那一行业塑造了波伊斯·布兰克斯地区的独特之处。围绕着欧洲科技中心的街道中间，有一条南北方向的通往布拉斯·德·康特勒运河的主道，决定着邻近街道的布局。运河和一些防渗水沟的两边种植着成排的白蜡树，一种让人能想起昔日沼泽地的树木。该设计实现了对径流的有效管理，方法是根据"道路花园"的理念，在德勒山谷对面建造一个水平面较低的花园，在园中种植有蓄洪特性且不会破坏环境协调性的植物，以汇聚附近的流水。

Omnipresent water directs the urban composition　无处不在的水影响着城市的构成

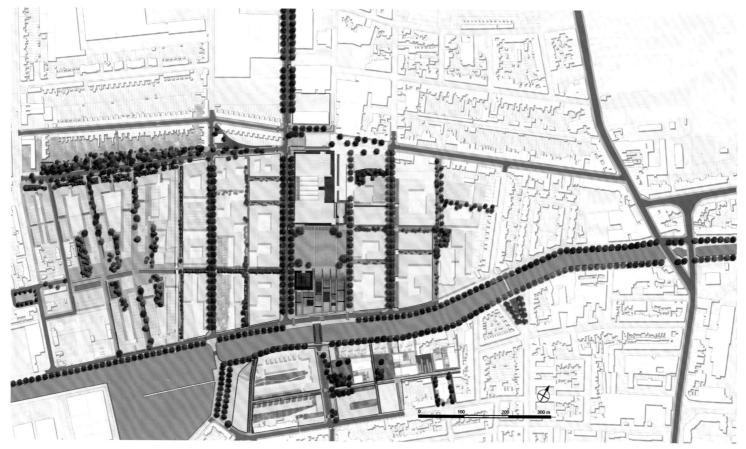

Facing the former cotton mill, the heart of the 'hydro-urban' process crystalizes in the water garden. Its role is above all technical because it collects, regulates, and purifies the runoff water through a process of phyto-purification. Its function as a storm basin gives way to that of the ornamental garden, punctuated by rooms and suspended passageways over the vegetation of aquatic environments that benefit the entire neighborhood. The installation of some hosts, such as ducks, moorhens, dragonflies, and other odonates, are indicators of its good ecological health, punctuated by rains.

In 2009 this project was awarded distinction in the eco-neighborhood category on the theme of water. It also won the 2010 prize for urban development. It was one of 13 accredited national eco-neighborhoods, a distinction given in 2013 by the Ministry of Ecology.

在朝向过去的棉花加工厂位置,"水城"的核心设备呈放散状分布在整个花园中。该设计主要是一项技术性设施,其作用是通过植物的作用收集、调节和净化径流水。"水城"内零星点缀的通路和房屋都设在水生植物的上方,那些水生植物改善了周边的生态环境质量。在暴风雨来临时,"水城"作为一个蓄洪设施,将接纳来自周围装饰性花园的流水。雨后,花园里不时前来造访的鸭子、雷鸟、蜻蜓等客人,成为该处生态健康水平的指示器。

此方案荣获法国 2009 年环保街区设计水主题的杰出方案奖,也获得了 2010 年的城市整治首奖。上德勒河岸是 13 个法国国家认证生态区之一,该荣誉由法国生态部在 2013 年授予。

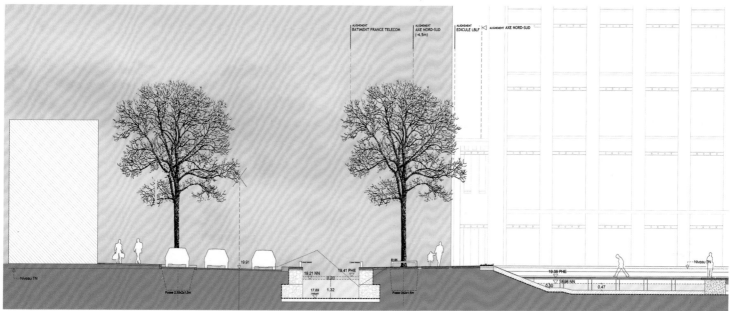

Water regains a central position in the public space : 水在公共空间中居于主导地位

Facing the old cotton mills, the heart of the 'hydro-urban' process crystalizes in the water garden 01 面对旧绵纺厂的水上花园使得"城市水化"过程初步成形

A space for phytoremediation space, the water garden is a place of leisure　02　水上花园可以起到以植物净化水质和美化空间的双重作用
Wet meadow and water garden in front of the façade of Euratechnologie　03　欧洲数码技术中心对面的湿地草甸和水上花园
A vocabulary borrowed from the industrial memory of the site　04-05　空间的形式语汇借自当地历史上的工业文化
The water garden provides biodiversity　06　提供生物多样性的水上花园

The management of runoff water constructs the urban project ┊ 城市地表径流水管理

- Hydrographie principale - Canal de la Haute Deûle
- Canaux en eau
- Noues végétalisées
- Jardins en creux
- Jardin d'eau
- Stockage superficiel enterré
- Parcs - Jardins

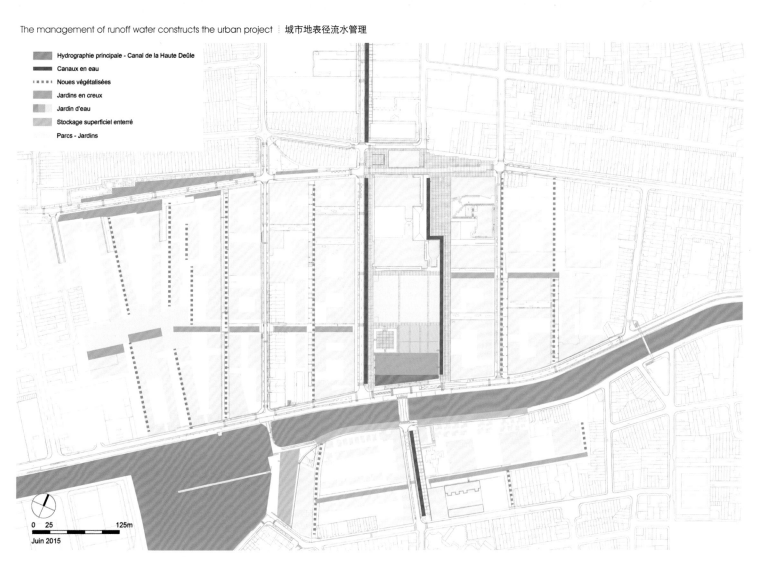

0 25 125m
Juin 2015

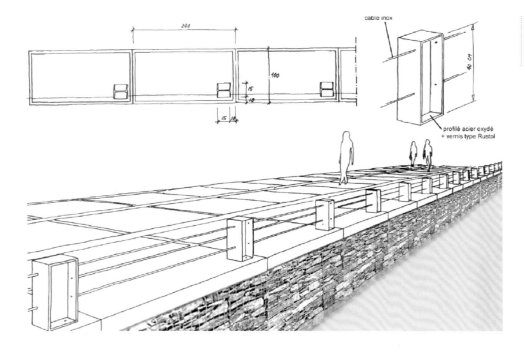

Sketch for a danger warning system at the banks of the canal

运河岸示警装置

Impromptu swim in a rainwater collection canal 07 游客一时兴起在收集雨水的运河中游泳
The canals orientate the promenade towards the Canal of the Deûle. 08 一系列水渠引领人们的漫步走向德勒运河。
At the end, the new residences of the Île des Bois Blancs 运河尽头是新建的"白色森林岛屿"住宅区
The parvis of Euratechnologie slopes towards the canal crossed by 09 欧洲数码科技中心的前庭广场朝向运河缓缓下降，
a metal passageway 运河上架设了金属过桥，供人通行

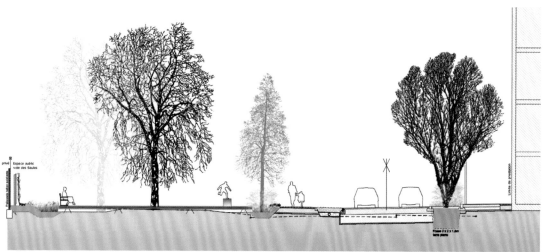

Sunken garden and big trees on the scale of the Deûle Valley

德勒河谷范围内的
下沉式花园和大树

The newly outlined avenue of willows exploits the spectacular vegetation of the ancient ditches of the Lomme marsh
Sunken gardens and riprap of black Flanders limestone secures the water management
At the foot of the old willows, the sunken garden, oriented east-west
Willow, alder, ash, and grasses of the cool environments constitute the new garden-street

10 新近开设的柳树大道，重新将昔日洛姆沼泽中沿着水渠种植的壮观植物加以利用
11 下沉式花园和法兰德斯石灰岩的设置确保了街区中雨水管理的效益
12 老柳树脚下东西走向的下沉式花园
13 适合寒冷地带生长的柳树，与赤杨、白蜡树和青草构成了这个新建的街道花园的主要景观

At the heart of the new residential buildings, paths with porous surfacing	14	在新居民建筑的中心地带，铺设的是高渗透性表面的道路
A new domestic and permeable landscape	15	一个具有渗水性的新居家环境景观
Ditch, elder tree and grasses become blocks of freshness	16	水沟、老树和草地为街区增加了生机

Ditch planted with vegetation, metal passageways, and 17　绿化生态沟渠、金属过道和预制水泥的水道头，
headworks in prefabricated concrete reveal this collected water　　陈述着雨水收集的故事
Ditch planted with vegetation; new urban water resource 18　绿化生态沟渠让水成为崭新的城市资源

L'Anton & Associés

Le Havre City Entrance
勒阿弗尔城市入口

Location｜地点
Le Havre, France
Completion date｜完工日期
2007–2018
Area｜面积
20 ha
Client｜业主
Ville du Havre
Contracted partner｜合作事务所
Infra Services, CDVia, Ingédia, F. Franjou
Photo credits｜图片版权
L'Anton & Associés

The Churchill and Leningrad Boulevards redevelopment project presents a major challenge for Le Havre. Along almost 2.5 kilometers, this road distributes the entire city: the port, the southern neighborhoods, and the town center. The road runs alongside sectors that are ripe for change and serves the new developments that structure the urban area. Churchill and Leningrad Boulevards run east to west, relatively coherently, with the parking lanes flanking them to the north and south. The project currently under development maximizes what is already there, while transforming its image and function.

Water purification in Le Havre is unified, so much so that the used water is diluted by the relatively clean rainwater. The project proposes to entirely reinfiltrate the rainwater collected on the 20 hectares of redeveloped public spaces. The particularly flat site (floodplains of the Seine river) makes a gravity-fed purification system by pipes impossible. A subhorizontal ditch is created along the roadway, collecting the rainwater from the pavement and protecting a cycling and walking path, created to the south of the road. The mini-underground pipes are not demolished but transformed into works for the storing and recovery of rainwater. This park also has the function of implementing a system for stormwater harvesting and for the soft cleaning of the urban runoff.

In Le Havre, a maritime and river city, the seaside atmosphere is noticeable everywhere (harbour basins, beach, and so forth), while that of the river has become less visible. The project aims to restore to the entrance to the city the estuary ambiance, the meeting point between the river and the sea, the brackish waters, both silty and salty, the willows, the reeds, the seaweed, and the seagulls.

In 2011, the project received First Place in the National Competition for Entrances to the City.

对勒阿弗尔市来说，丘吉尔和列宁格勒林荫大道的整治计划是一个相当重要的挑战。这条近2.5千米长的道路连接了港口、南部街区和市中心。然而此条道路呈现出完全属于机动车的面貌，位于交叉路口下方供汽车穿越的小型地下通道、高架桥甚至立交桥一个接着一个。道路沿途经过城市中一些不断变化的街区，并为在整个城乡区域占有结构作用的若干发展中心提供服务。丘吉尔和列宁格勒林荫大道的路向是从东到西，其南北方向都有停车场紧邻，随着项目的实施，原有设施的功能必将被进一步发挥到最大，但原有设施的外观和功能也必然发生变化。

勒阿弗尔的水净化系统相当发达，以至于可以用相对干净的雨水在过滤后用做饮用水。设计规划在约8公顷地区的公共空间范围内，进一步强化雨水收集过滤系统的效能。在特别平坦的区域（主要是塞纳河河床地区）设置一个利用引力的管道输水净化系统。道路南边修建了辅助性的平行水沟，用以收集碎石路上排下的雨水，并起到保护自行车道和步道的作用。小型的地下管道未被拆除，被改造成用于储存和回收雨水的工程系统。该园区还具有雨水收集和初步处理城市废水的功能。

在勒阿弗尔这个海滨及河岸城市，到处都能感受到海洋的气息（港口、海滩等），然而关于河流的元素却极其稀少。本方案致力于在城市入口处重新营造小港湾的气氛，让海洋与河流在此相遇，届时，与咸海水相伴而来的，除了混浊的淤泥、柳树、芦苇、海草，还将有在天空翱翔的海鸥。

2011年，该项目设计赢得了法国国家城市入口设计大赛第一名。

Principle of water purification｜水净化主要流程

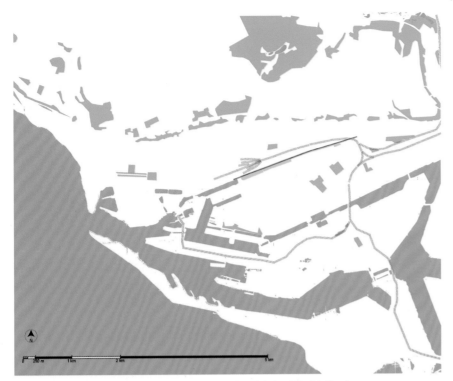

The entrance to the city and to the port of Le Havre : 勒阿弗尔市和勒阿弗尔港口入口

View of the whole project　01　项目全景
Big water storage ditch　02　入口大道上的主要生态贮流渠

The system of rainwater purification and reinfiltration | 雨水净化和重新渗入土壤系统

The subhorizontal ditches　03-05　呈水平状的生态沟渠
The future park of the Docks　06　未来的码头公园

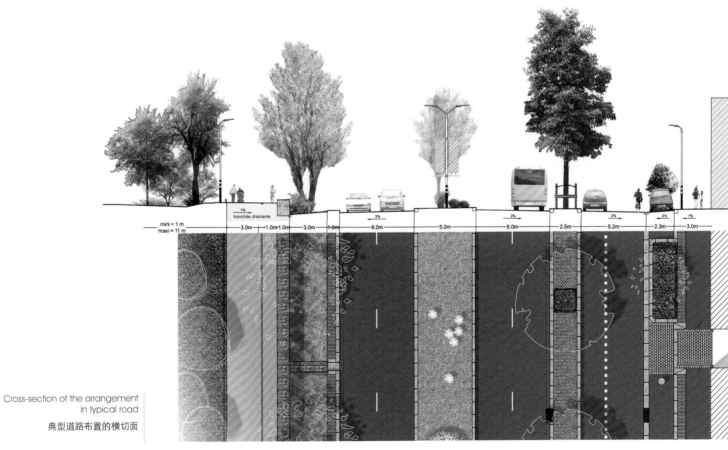

Cross-section of the arrangement in typical road
典型道路布置的横切面

View of the whole development 07 项目开发全景
Subhorizontal ditches 08 呈水平状的生态沟渠
Promenade along the subhorizontal ditch 09 沿着水平状生态沟渠分布的散步道

Saint-Martin-d'Hères, a town on the eastern periphery of Grenoble, has undergone heavy development in the last years. Lucie Aubrac Square accommodates the town's major facilities and was renovated in 2007 when the Line D Tramway was introduced. While it provided an opportunity to enrich the uses of this fragmented space and make it more coherent, the challenge has been the collection and infiltration of rainwater in this heavily frequented area.

The goal of managing rainwater on site in pedestrian areas can be seen in the first sketches of the project's conception. At the launch of the project in 2005, this type of operation had not really been developed before at a town level. In this project, rainwater management is segmented in many collection points. Water is channeled towards a group of vegetal islands that receive and then infiltrate the water from pedestrian surfaces. The vegetal islands, of varying sizes, abundantly planted with willows and poplars, draw attention to the area and confer on it the image of a verdant garden square.

These rain gardens are designed to gather the runoff water from the square as well as from the roofs of certain buildings. The water from the roads is also sent to this network. In case of heavy rain, when the infiltration is not sufficient, the ground runoff is thus managed by the vegetal islands. The development project of Lucie Aubrac Square has been able to take advantage of the multiple constraints of the place while creating a quality neighborhood environment through this garden square.

In Situ

Lucie Aubrac Square
露西·奥布莱克广场

Location：地点
Saint-Martin-d'Hères, France
Completion date：完工日期
2007
Area：面积
0.8 ha
Client：业主
GPV Grenoble, Ville de Saint-Martin-d'Hères
Contracted partner：合作事务所
E2CA
Photo credits：图片版权
In Situ / Emmanuel Jalbert

圣马丁厄尔是一个小镇，位于法国格勒诺布尔市东部边缘地区，在过去的几年中经历了迅猛的发展。露西·奥布莱克广场是这座小镇的主要基础设施。广场改造项目的提出是由于市区D号线电车在2007年通到了该处。项目设计的真正挑战是在这个降雨频繁而且降雨量很大的地区解决如何收集和涵养雨水问题。项目还提供了一个机会，就是进一步丰富广场的用途，并借此整合小镇破碎的空间。

在项目的第一个草图中，可以看到上面列出了行人区雨水管理的目标。2005年这个项目启动之际，这一类型的规划还没有在全小镇范围内被真正地实施。在这个项目中，雨水管理被设计成在许多收集点分散进行。具体来说，降水被导向一组种满植物的绿色"岛屿"园圃，积蓄在那里并经步道表面渗透进地下。种植植物的"岛屿"大小不等，上面的树木主要是柳树和杨树，打造出一个翠绿的花园广场，吸引着游客不断来访。

设计这些雨水园圃的目的是从广场上以及建筑物的屋顶收集径流的雨水。落在道路上的雨水最后也会被输送进到这个网络中。在下大雨的情况下，当街道地面渗透作用无法及时消化掉过多的雨水时，植物"岛屿"开始发挥作用，地面径流即交由"岛屿"来吸收。露西·奥布莱克广场开发项目的实施，通过建造一个花园广场提高了邻近地区环境的品质，目前已经让这个存在诸多不利因素的小镇因此而受益。

Master plan：总体规划图

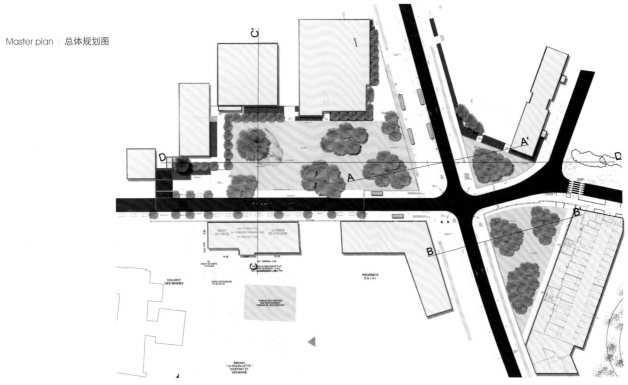

The landscaping of the garden-islands (first year), planting of 'living' 01 第一年的"岛屿"花园，周围是
fences in braided willow that marks the identity of the place and 柳条编制的"活"篱笆，具有提升场所特色和
protects the young plants 保护幼小植物的作用
The islands of vegetation collect and infiltrate rainwater 02 收集和渗透雨水的植物"岛屿"
Current development of the vegetation of the garden-islands 03 植物"岛屿"的现状

Illustration of the principal of rainwater recuperation presented on the site 04	项目雨水回收系统主要流程图解
Interior view of a little island: plantings of perennials and a gravel bed to 05	"岛屿"内部景象，可以见到用于
facilitate rainwater infiltration	吸收雨水的沙土圃地和多年生植物
The garden-islands create micro-sites, offering more intimate spaces 06-07	"岛屿"在这个大广场上创造的微观环境为人们提供了
in the centre of this large esplanade	更多亲密接触大自然的机会

Paths cross the garden-islands 08 穿过"岛屿"的道路
The islands provide shade and coolness, an evident asset 09 提供阴凉的"岛屿"
in the urban environment 　　成为城市环境中的重要生态资源
The abundant vegetation evokes wet zones. The development 10 大量植物营造了潮湿的环境，
of this vegetation confirm the proper functioning of the islands 　　植物的繁茂确保了"岛屿"功能的正常实施

Atelier de l'île – Bernard Cavalié Paysagistes

Three Rivers Walkway
三河林荫道

Location｜地点
Stains, France
Completion date｜完工日期
2008
Area｜面积
1.8 ha
Client｜业主
Plaine Commune 93
Contracted partner｜合作事务所
Ingema – Quetzal ingénierie
Photo credits｜图片版权
Atelier de l'île / Isabelle Otto

Three Rivers Walkway is first and foremost a public space acting as an interface between Georges Valbon park, the existing neighborhoods, and the new residential fabric. Along almost 700 meters, it offers the inhabitants new outdoor spaces close to their homes and creates links: places for relaxation and recreation, soft links between neighborhoods (access to facilities), and a walk to the departmental park. A floodable space, it allows for the stormwater management of the Three Rivers neighborhood.

Heavy rain is not seen as an accident to be evacuated as quickly as possible, but as an element to enliven the site: the spaces created are flooded at the mercy of the rains, each in its own way. Grass-covered swales are not the only possible way of laying out these floodable spaces, and they have been treated differently according to their immediate context and the possible ways they will be used. Thus, along the soft link, large meadows, wetland zones, or floodable hard-surfaced squares have been designed as spaces close to nature, with different environments and uses.

Not wishing to store rainwater to the detriment of the users of the walkway, the landscape architects had to find the best way to adapt the technical solutions to the intended uses and environments. Thus, the spaces that are most heavily frequented (the lawn, the wave square), with soft forms, flood less often and dry out very quickly, while the low point of the site has been deeply dug out and partially waterproofed in order to retain the water for longer. Playing fields, planted water meadows, hard surface squares, and wetland zones are primarily spaces for leisure and use. Evolving according to the amount of rainfall, they allow people to rediscover a contact with nature even in a dense urban environment, and diversify the presence of water in the city.

三河林荫道是介于乔治斯·瓦尔邦公园、现有街区和新住宅区之间的公共空间。它在近乎700米的长度上为周边居民提供邻近的交流空间和通行空间。它是休闲游憩场所、街区之间的软性连接（可通往各种公共设施），也是通向省级公园的散步道。这片可吸收洪泛的区域使三河街区的雨水得以获得较好的控制和管理。

降雨在此方案中并不被视为是需要尽快排除的意外事件，反而被看作是基地内的活力元素：每个空间将随着降雨情况而以不同方式被雨水淹盖。设置草坪斜沟并不是受淹区域唯一的处理方式，设计师根据不同空间的环境背景和潜在的使用功能而采取不同的处理方式。林荫道上的各个空间，大草坪、湿地或可吸收洪泛的硬质小广场，都被设计成贴近自然的、拥有不同使用功能和气氛的空间。

为了避免雨水的淤积不利于林荫道的使用，方案所采用的技术措施需要适合空间环境和实用需求。那些人们经常使用的场所（草地、波浪广场）地势变化平缓，偶尔会遭到水淹但又可以很快地排干；反之，某些低陷空间则被深入挖凿且局部防止渗水，以便可以更长久地保留雨水。游乐场、湿地草甸、硬质广场以及湿地主要是提供居民消遣娱乐的实用场所。这些场所随着降雨情况而产生变化，使生活在高密度城市环境中的人们也能重新与自然接触，并且认识水在城市中的不同表现形式。

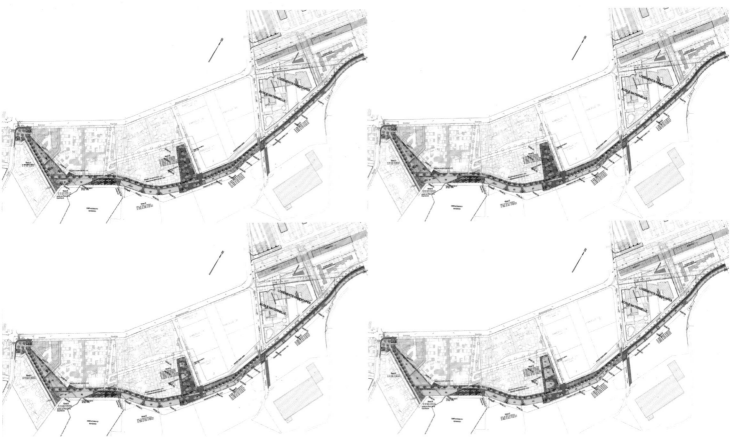

Schematic plans of the planted ditches filling with water according to the types of rainfall | 不同降雨量时绿化水沟功能图解

Views of the slightly elevated walkways and the ditches below　01-03　生态沟渠上地势稍高的步道即景

View of the entire walkway along the wet zone	04	湿地步道全景
The flood plain beautifies the new entrance towards Georges Valbon Park	05	洪泛区的整治美化了朝向乔治斯·瓦尔邦公园的新出入口
Play area on the flood plain	06	洪泛区的游乐空间

| Schematic perspective of the square surfaced in brick as it fills with water | 铺砖广场在被雨水注满后的情况图解 |

Flood plain along the walkway　07　步道旁边的洪泛区
View of the entire square surfaced in brick　08　铺砖广场全景
The floodable depressions in the square surfaced in brick　09　铺砖广场上吸收洪泛的洼地

Agence Territoires

Large Meadow
大草原

Location｜地点
Sermange, France
Completion date｜完工日期
2008
Area｜面积
0.63 ha
Client｜业主
Commune de Sermange
Photo credits｜图片版权
Nicolas Watelfaugle

The village of Sermange, located in the department of Jura, is built around an extensive lawned area called the Large Meadow. This area, long used for markets, drains the runoff rainwater from part of the village. A fountain occupies the lower part of the site.

The project makes a simple and readable statement in attempting to construct the limits of the Large Meadow by aligned plantations of ash, and by a wooden path that plays with the topography in order to enhance the subtlety of its undulations. By not intervening on its center, the project preserves an empty space suitable for multiple communal activities and maintains the natural hydrography of the place. The rainwater that runs off at ground level joins the water that progresses underground, at a new basin. The soil keeps its floodable character and forms a playground and educational space for the village school.

In 2010, the project received the National Prize of Urban Development in the category of towns of fewer than 10 000 inhabitants.

位于汝拉山区的塞尔芒吉村庄是围绕着一片名为"大草原"的广阔草地而发展起来的。此空间长期以来一直作为市集用地使用，并且汇集了从村庄范围内流过的一部分径流，在草地的低洼处形成一个喷泉。

该项目规划的实施，要求有关方面发布一个简单易懂的声明，以禁止居民对"大草原"的烧荒造田，为此将在"大草原"周边建造一条木板路，以界定"大草原"的稳定地理边界。通过保持中心地区原生态的策略，设计在"大草原"内部保留了一片适合的空间，以便周边民众在那里开展多种公共活动，同时又保持了地方自然水文的一贯性。在地面上流动的雨水以及地下水被引入一个蓄水盆地，使得"大草原"的土壤依旧保持着防洪蓄水的特性，同时还为当地开辟出一个可用于教学的乡村学校操场。

2010 年，该项目获得了法国国家城市发展奖（居民在一万人以下小城镇类项目）。

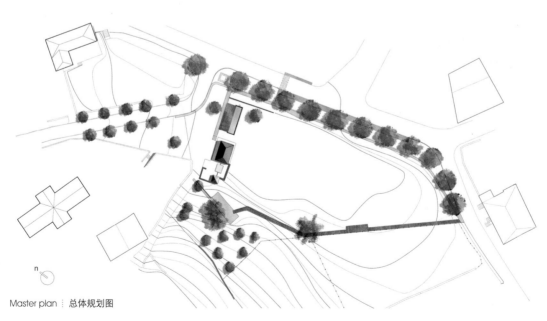

Master plan : 总体规划图

General view of the large meadow　01　大草原全景
The wooden walkway, new pedestrian path in the village　02　村内的新木制人行步道

Detail of the new pond ┆ 新池塘细部

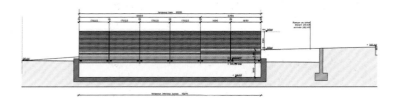
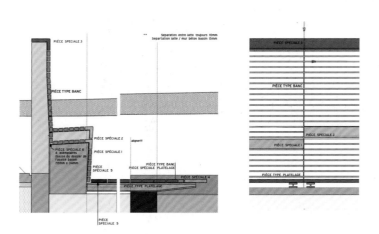

The large meadow: the ancient fountain washhouse towards which rainwaters are concentrated	03	大草原内的古代供水点和洗衣池，成为雨水集中的空间
The ancient fountain washhouse is lengthened by a new pond as well as a large deck and a bench	04	被扩大后的供水点和洗衣池，增加了一个新池塘、一大面甲板和一张长椅
The new pond gathers the water of the fountain and gives it back to the wet soil of the large meadow	05–06	新池塘汇聚供水点和洗衣池流出的水，再将其用来灌溉大草原的土地

The large meadow, game fields at the centre of the village　07-11　大草原成为村落中的休闲游乐场所

147

Agence Territoires

New TGV Station of Belfort-Montbéliard
贝尔福-蒙贝利亚尔高速火车新车站

Location 地点
Meroux, France
Completion date 完工日期
2011
Area 面积
5.1 ha
Client 业主
SNCF
Contracted partner 合作事务所
DAAB, AREP, D'Ascia, RFR, Le Point Lumineux
Photo credits 图片版权
Nicolas Waltefaugle

A train station is a functional space. Yet the new Belfort-Montbéliard station offers an exceptional experience, facing the blue Vosges mountains. This encounter between the world of movement and the stability of the geography of Franche-Comté shapes a project where the relationship with nature is essential. The panoramas of this setting are protected, revealed, and amplified. Lookout points over the landscape are provided by constructed frameworks along the route of train travellers from the parking lot to the station. The 1 200 parking spots are thus a pretext for an organisation of the space that reveals the geography of Franche-Comté and places people at the heart of this natural setting.

The system for managing runoff water establishes the natural quality of the project. Sensitive to the topography, the succession of ditches forms a chain of environments ranging from the driest to the wettest. The last ditch, which becomes a true garden both for the station users and for the local flora and fauna that it hosts, is ensconced in the thalweg of this valley. This garden is yet another way of transforming the project, which fundamentally is very technical, into a tool for the understanding of the territory.

The extensive network forms a system of rainwater collection, the water filling then pouring out from pool to pool. As it makes its way, the water is filtered a first time and feeds a wetland flora. Finally, the whole system feeds the principal reservoir that in turn purifies the water of the parking lot and forms the garden around which the functional environment of the train station is organized.

车站是功能性空间,然而面对孚日山脉山间河流的贝尔福-蒙贝利亚尔新车站却为人们提供了十分独特的体验。当火车世界的运动特性和弗朗什-孔泰地区风景的稳定性发生冲突时,一个旨在处理好与大自然之间紧密关系的崭新设计方案被促成了。设计确保车站基址位置的景观受到保护,而且被予以展示甚至强化。从停车场到火车站的沿路的基础设施中,设置了很多观景点,可以让旅行者看到前述风景。1,200个停车位的规划组织也致力于向旅行者展示弗朗什-孔泰地区的地形地貌,让人们置身于大自然的怀抱之中。

地面径流的管理设施建设遵循不改变景观的自然状态原则,通过对地形的精确利用,借助一系列山沟展示从干燥到潮湿的多样地理环境。最后一个山沟被改造成为可供车站旅客欣赏的景观花园,也是动植物汇聚的生态花园。这个原本技术性很强的项目被转化为人们理解大地景观的工具。

雨水收集系统的网络很广,雨水通过这个系统充满一个又一个水池。当系统发生作用时,雨水在被过滤后,用于灌溉湿地植物。最终,雨水通过系统汇聚到主水库中,停车场上排出的废水也会被输送到水库中,并在此得到净化。水库中的水还可以用于灌溉车站周边的功能花园。

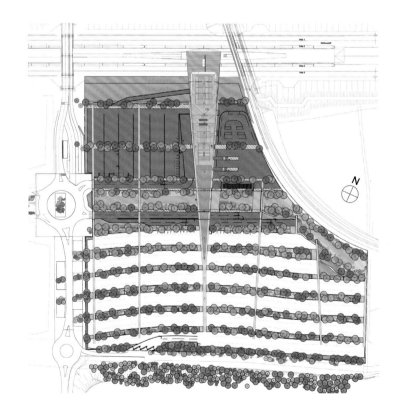

Master plan:
the outline of the ditches follows the curves of the levels. Rainwater travels from one ditch to the other down to the central thalweg

总体规划:
水沟沿着地势高低设置,雨水从一条水沟流动到另一条水沟,最后抵达位于中央位置的水位最低的水沟

The functioning of the high-speed train station is organized around the thalweg, the final destination of the rainwater. It constitutes at once the garden of the train station, around which pedestrians, cyclists and vehicles gravitate

01-02 高速火车站的功能性空间围绕着一道生态山沟而配置。此山沟是雨水收集的最终目的地，同时也成为车站花园，周边设置了步道、自行车道和机动车道

The north-south pedestrian path
南北走向的步道

The thalweg of the train station follows and respects the outline of the original thalweg. It brings together a rich flora that grows well with the gradients of variable humidity and that the pedestrian and vehicles cross and observe

03-05 车站的生态山沟循着原有的山沟位置而开发，山沟带来了大量适应不同坡度湿度条件的植物。在跨越其上的步道或车道可以观察到这些植物

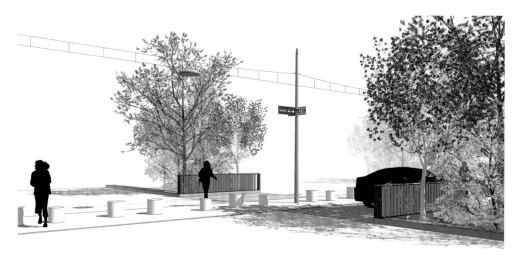

The north-south pedestrian path cuts through the ditches and the thalweg | 南北走向、穿过多条水沟的步道

151

The landscaping project of the train station aims at integrating this facility in the geography that accommodates it: marrying the incline of the terrain, lengthening the forest edge and orientating the views towards the principal points of the area, the orchid hill and the Vosges ridge

06—09 火车站的景观设计致力于将车站设施和周围的地理环境整合为一体：顺应地形坡度、延伸森林的边界地带、将人们的视野导向当地的几个重要景点——兰花丘陵和佛日山脉

Atelier Villes & Paysages

Zénith Concert Hall Public Realm
天顶音乐厅室外空间

Location | 地点
Strasbourg, France
Completion date | 完工日期
2008
Area | 面积
30 ha
Client | 业主
Strasbourg Eurométropole
Contracted partner | 合作事务所
Agence Becard & Palay, Egis, L'Atelier Lumière / Pierre Nègre
Photo credits | 图片版权
Balloïde Photos (n°01–08), Atelier Villes & Paysages / Emmanuel Moro (n°09–12)

The main challenge in the development of the outdoor spaces of the Strausbourg Zénith Concert Hall was to create a parking lot capable of containing 3 500 light vehicles, as well as motorcycles, bicycles, and buses. The treatment of runoff water was the object of particular attention: infiltration being impossible, the idea was to gather all of the water through transversal drainage ditches, releasing it into a series of water retention basins, the whole system being entirely sealed.

The sunken construction of the central drainage ditch allows the collecting of rainwater originating from alveoles. On rainy days, it forms a wet landscape, in contrast to the landscape of dry days. The ditch also helps to slow down the flow, before infiltration, at the end of the water route. This drainage ditch is specifically placed in a longitudinal direction and is segmented by crossing paths. It is planted in a natural way with a range of plants from wet environments and its aborescent stratum is composed of coppiced trees and saplings planted randomly.

The creation of these drainage ditches has made it possible to propose a landscape rich in plant diversity and to alternate the parking places. As for the ponds, they mark the principal strategic axis and create a breathing space at the heart of the 'park' in the direction of the Zénith.

斯特拉斯堡天顶音乐厅室外空间项目主要包括一个能容纳3,500个停车位的停车场的建设，停车场可用于轻型汽车、摩托车、自行车和公共汽车的停放。雨水处理是项目特别关注的对象——既然在当地让雨水向土壤渗透是不可能的，可行的方法就是收集所有的水，通过横向排水沟释放到一系列的水的涵养低地，这要求整个系统是完全不透水的。

相对水平高度较低的主排水沟配备有蜂窝状的孔穴结构以汇聚雨水。比起晴朗的天气，这里在下雨天愈发呈现出一片湿地景观的模样。水沟也延缓了水的流动速度，以便在水路的终点对水进行净化处理。水沟被特别设计成纵向分布，中间不时切入小路。项目区域内以自然方式种有一系列适合湿润环境的植物，到处是随意栽植的矮灌木和小树。

这些排水沟设计使得在项目区域内栽植多种植物、发展景观的丰富多样性成为可能，也使得让植物和停车位交错分布成为可能。池塘的主要作用是从规模总体结构的角度，在属于规划项目区中心的天顶音乐厅对面创建一个可供人休憩的空间。

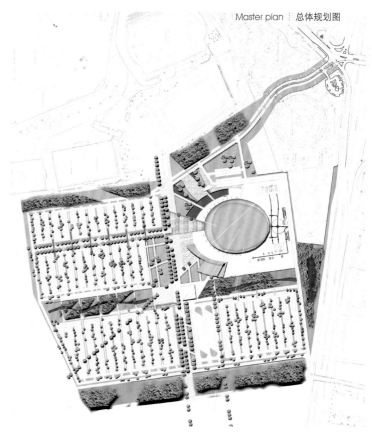

Master plan : 总体规划图

Integration of vegetation in the parking areas　01　停车场区域的植被整合
Aerial view of the entire parking lot beautifying the Zénith　02　绿化停车场鸟瞰景观：此停车场的整治美化了天顶音乐厅的整体环境

Layering of vegetation　03　具有不同高度层级的植物景观
Infiltration ditches　04–05　渗水生态沟渠
Vegetal and mineral vocabulary borrowed from the urban park　06　借用城市公园设计所使用的软质与硬质元素语汇构成设计理念的关键字

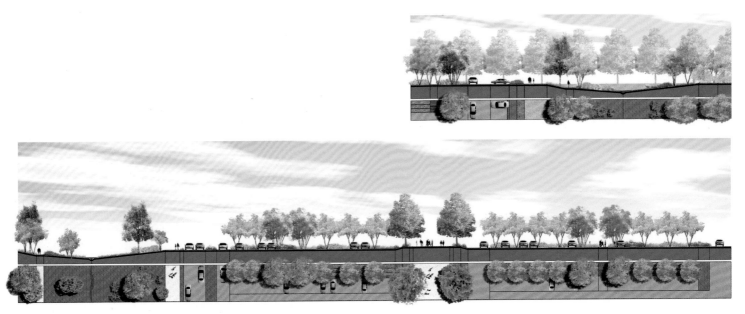

Basic sections: 原则性剖面图：
integration of basins and parking areas　生态贮流沟和停车空间的交错配置与融合

| Water-tight basins in series: the collection of waters coming from the ditches | 07 | 不透水的系列蓄水池：汇集来自生态沟渠的水体 |
| Paths that cross the alignment of ditches | 08 | 穿越一系列生态沟渠的小径 |

Vegetation that accompanies the ditches 09 生态沟渠边的植物
The transversal ditches create vegetal spaces: 10 停车位间的横向小型生态沟渠
parking spots 创造出植物的生长空间
The exterior developments of the Zenith break 11 天顶音乐厅的外部空间设计颠覆了停车场的
with the orthogonal logic of the parking lot 矩形空间设计理念
Terminal pond after heavy rains 12 大雨后的终点蓄水池

Rainwater Parks

雨水公园

Urbicus

Prés-Devant Suburb
布莱-得旺市郊

Location｜地点
Chalon-sur-Saône, France
Completion date｜完工日期
2013
Area｜面积
12 ha
Client｜业主
SEM de Bourgogne
Contracted partner｜合作事务所
PMM, Cabinet Reilé
Photo credits｜图片版权
Urbicus

The construction of the new Chalon-sur-Saône hospital, together with the completion of the bypass, necessitated retroactive thinking about the whole suburb of Prés-Devant. The issues at stake were the redefinition of public spaces, traffic flow, the development of a medical hub, the reduction of uncontrolled landfill, the fight against invasive plants, and flood management for the River Thalie, all within a very limited project budget.

The old departmental road became a walkway through the park; waste landfill created a new topography; the Thalie wetlands were preserved; and the public park became floodable. Using the constraints and residual problems as the basis for a project generated a new plain and positive landscape.

The necessity of landscaping the soil to accommodate rainwater retention structures guided the development of the park. The circulation of water brings life to all the various areas: the water from the roofs of the hospital building partly feeds the permanent pond and finds an outlet in the Thalie Valley. Inversely, in the case of spikes in the height of the river, the water is sent towards the park and can expand into the wet zones and into one of the ponds.

索恩河畔沙隆新医院的设立和城市环线的完成，迫使人们重新对其所属布莱-得旺市的郊区的发展进行一个追溯既往的整体性思考，包括其公共空间的重新定位、交通流量的规划、医学设施的发展、被随意丢弃之垃圾废物的回收、对蔓延性植物的防范、对塔利河涨潮的控制，以及对项目有限预算的管理。

从前的省级公路变成了公园的林荫道，废物的集中地产生了新的地貌，塔利河的湿地形成了花园，而公园则成为可吸收洪泛的地区。各种限制和未被解决的问题成为项目设计的新动机，创造了一些平凡无奇却起到了正面作用的新景观。

园林绿化的必要性在于园林对水土保持所发挥的作用，这一点是公园发展的领导思想。水的循环给所有区域都带来了活力——从医院建筑的屋顶落下的雨水为院落中的永久池塘提供了部分水源，并最终在塔利河畔找到一个入口。相反的，在河流水位高企的情况下，河里的水可以被输送回公园，疏散到湿地和各个池塘中。

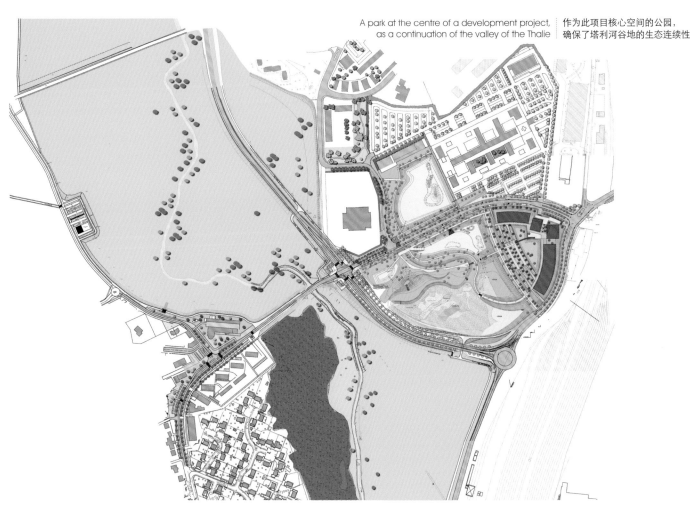

A park at the centre of a development project, as a continuation of the valley of the Thalie

作为此项目核心空间的公园，确保了塔利河谷地的生态连续性

The regulatory constraints linked to the risk of flooding guide the levelling of the park | 根据防洪法规的要求而进行的公园整地工程

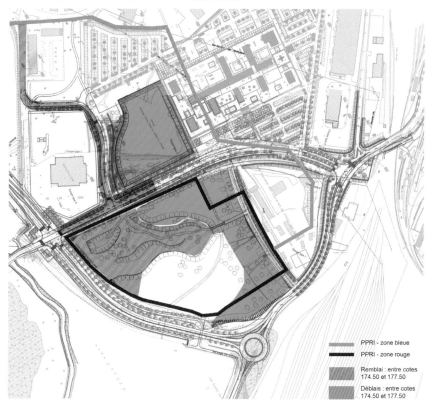

PPRI - zone bleue
PPRI - zone rouge
Remblai : entre cotes 174.50 et 177.50
Déblais : entre cotes 174.50 et 177.50

The hospital building framed and enhanced by the ponds　01　为池塘所界定和烘托的医院建筑
The ponds are integrated into the landscape of the existing wet zone　02　池塘被整合进现有湿地景观

Principal of feeding water to the ponds and of the direction of the water flow in normal periods

平常状况下池塘用水的供应和水流方向

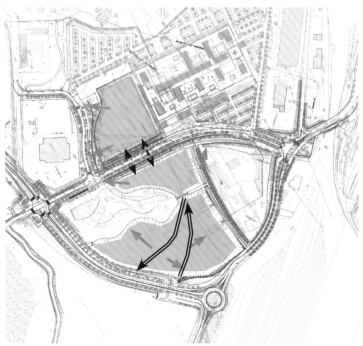

Principal of the circulation of water during rises in the river and of the retention zone

遭遇洪峰期间河流和蓄水区域的水循环过程

The low areas of the park accommodate the rises of the Thalie river during the winter	03-04	公园的地势较低区域可应对冬季塔利河水位的增加
A park with a natural ambiance	05	天然氛围的公园
Details of the landscaping design	06-07	项目细部

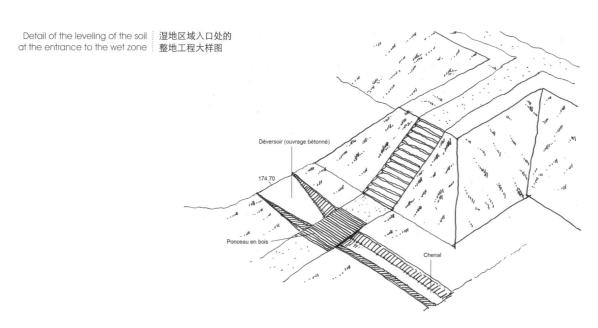

Detail of the leveling of the soil at the entrance to the wet zone | 湿地区域入口处的整地工程大样图

A very floodable park 08-09 一个蓄洪能力很强的公园
The bus lane is a thoroughfare of the park, 10 公共汽车道是公园的主道，路边的
bordered by ditches supplying the ponds with water   生态沟渠将收集的雨水输向池塘

Situated on former agricultural land, Mount Évrin Park forms the backbone of a new, yet-to-be-built neighborhood. Stretching over 1.8 kilometers in length, the park provides a link between the old town, on a hillside above the Marne River, and the infrastructure and services of the new city of Marne-la-Vallée, developed on the plateau.

Its atmosphere is enhanced by the rural landscapes of the surrounding countryside: meadows, hedgerows, small woods, and a broad wetland resulting from the creation of a storm basin. Integrated into the heart of the residential area, the Grand Orchard offers the most garden-like atmosphere in the park.

The storm basin makes possible the collection of rainwater from the bordering urban area and from the eco-neighborhood on its fringes. A floodable meadow, traversed by irrigation ditches and punctuated by pollarded willows, has helped to create a wet zone that looks like a water garden. The natural texture of the wet zone permits it to integrate the technical structures necessary to the functioning of this hydraulic system.

Urbicus

Mount Évrin Park
艾夫兰山公园

Location : 地点
Montévrain, France
Completion date : 完工日期
2005–2018
Area : 面积
20 ha
Client : 业主
EPA Marne, Ville de Montévrain
Contracted partner : 合作事务所
Prolog Hydrologie, ATPI Infra, Zoom Écologue, Pixelum Lumière, Gras Miroux Architectes Associés
Photo credits : 图片版权
Urbicus

位于昔日农业用地上的艾夫兰山公园是即将建成的新街区的脊柱式设施。公园延伸1.8千米之长，连接了坐落在马恩河畔小山丘上的古老村镇和在高原上刚刚开发的马恩河谷新城的公共服务设施。

公园享有周边乡村特有的田园景观：草地、灌木围篱、小树林，以及一块将用于建立雨季储水池的大型湿地。大果园位于住宅区正中央，呈现出非常田园化的氛围。

针对暴风雨的蓄洪盆地的设立，使得在该处容纳邻近的城市地区和其边缘的生态社区释放的雨水变成可能。蓄洪草地中灌溉沟渠纵横交错，不时出现被修剪过的柳树，使得园区内保持着潮湿的环境，看起来就像是一个水上花园。这个水世界的周边自然环境也允许它集成要实现其功能所必需的技术结构。

The eco-neighborhood blocks are connected to the park | 生态街区与公园连接在一起

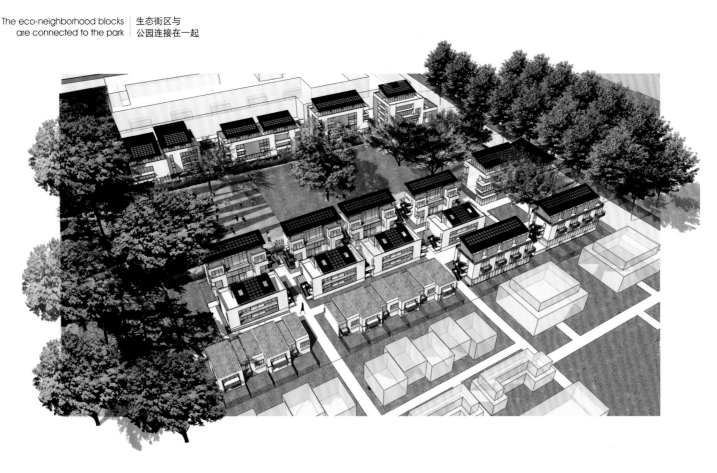

The park as the framework of the new Montévrain neighborhood

公园重新整合了
艾夫兰山公园邻近地区

The 'big path' crossing all the atmospheres of the park 01 穿越基地的"大道"整合了公园的各种生态环境
The ditches of the big orchard organise the space 02 大果园的生态沟渠起到了组织空间和
and also collect runoff water 收集雨水的功能

The wet zone of the former marl pit at the lowest point of the park collecting part of the runoff water | 从前的泥灰坑湿地区域在公园内地势最低，收集了部分地面径流

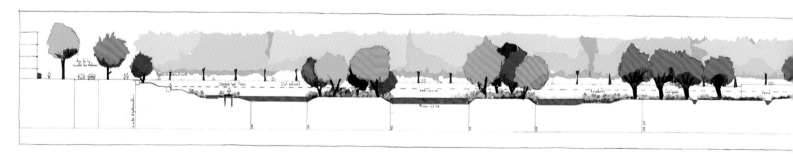

The pond of Charbonnière and its little islands conducive to the nest-building of birds | 查波涅尔池塘和中间的小岛，有利于小鸟在上面筑巢

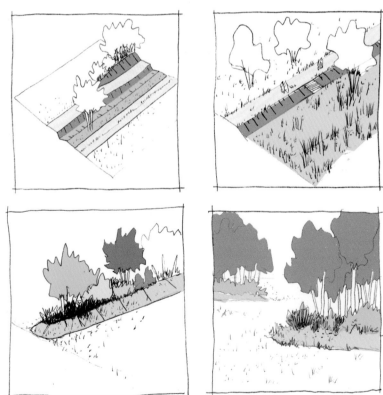

Different types of developments | 各种类型的方案

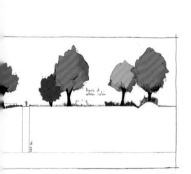

The meadow in the month of June, refuge for insects　03　六月间的大草地，成为昆虫的庇护地
The moving paths in the interior of the meadow　04　大草地中间的动人小径

The mineral rooms of the big orchard　05　大果园内的硬质地面空间
The green rooms created by a particular management around the furniture　06　设置有特制城市小品的绿色小沙龙
Details and furniture built to order　07–10　特别订制的座椅和地面处理细部

Atelier Ruelle

Glonnières Park
格隆涅尔公园

Location：地点
Le Mans, France
Completion date：完工日期
2012
Area：面积
7 ha
Client：业主
Le Mans Métropole
Contracted partner：合作事务所
Cabinet Bourgois
Photo credits：图片版权
Magdeleine Bonnamour

One of the levers of this project of urban renewal is the rehabilitation of the central park. Starting with the existing assets and reorganising the spaces near each building (access, parking, and so forth), the central space recovers all of its richness. The soil modeling is done a little at a time to limit earth-moving and to preserve the large trees.

The movement of earth has allowed the creation of ditches that reinforce the transition between the neighborhood and the park, situated at the center of the residences. A long ditch stretched into an arc of a circle reflects the geometry of the buildings and draws a limit, a filter, between the park and the nearby residential areas. Pontoons permit crossing the water zones and enhance the diversity of the pedestrian paths towards the interior of the park.

The quality of the developments plays between porous and less-porous soils according to the trees and their preservation or intensity of the use. At the foot of the services and to the right of the playgrounds, the width of the permeable joints narrows.

这一项目的改造更新动力来自中央公园的重建。从邻近本项目的现有资产和空间重组的情况来看（道路、停车……），中央公园已经恢复其一度出让的所有财富。例如，园中用于限制搬运土壤和保护大树的土壤模型一次性地被建造了很多。

土壤的存在使得挖掘沟渠成为可能，这些沟渠加强了邻近地区和被周围居民区包围的公园之间的界限。一条长沟蜿蜒前行，伸展到一座建筑的拱形结构下面，为长沟划定了界限。这座建筑位于公园和邻近的居民区之间，用于安装过滤设施。水面区域路段的浮桥设置，使得公园内的步行道路显得多姿多彩。

土壤渗透性强弱与否，跟树木的保护程度和分布密度有关。在公园下方、操场右边，具有渗透性的绿化区域渐渐变窄。

Master plan of the Glonnières neighborhood｜格隆涅尔街区总体规划图

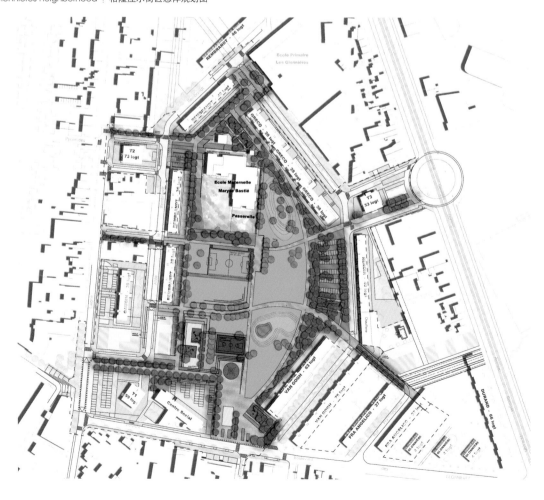

More porous soils, new layers of vegetation　01　整治方案为公园带来更多渗水性的地面和更多样的植物层次
Impermeable surface has been reduced to benefit the park　02　原本大量不渗水地面经过改造后面积被降至最低，有利于公园生态环境的发展

The big trees are preserved by a careful leveling 03 细心的整地工程后保留下来的大树
In the places of the most intense use, 04 在较为频繁使用的空间里，
the ground is duly less porous 地面的渗水性被适度地降低
A play on pavements on the theme of permeability 05 根据不同渗水需求与条件而搭配组成的地面处理

A big ditch delineates the border of the park　06　位于公园边缘的大型生态沟渠
An outline, a landscape... the ditch offers transitions　07　作为项目的结构线条和景观,此生态沟渠也成为不同区域间的过渡带

L'Anton & Associés

Park of the Savèze
拉萨维泽公园

Location | 地点
Saint-Herblain, France
Completion date | 完工日期
2015
Area | 面积
6 ha
Client | 业主
Ville de Saint-Herblain
Contracted partner | 合作事务所
Infra Service, Du§Ma, F. Franjou, R. Zumbiehl
Photo credits | 图片版权
L'Anton & Associés

The Sillon de Bretagne (Brittany Furrow) is a name that refers not only to a geological fault, but also to a huge building almost a kilometer long and 100 meters high, built at the beginning of the 1970s in Saint-Herblain, in the suburbs of Nantes. At that time, to the south of the building, a six-hectare park on the urban fringes was laid out at a minimum. Consequently, this little-used park offers zero or even negative biodiversity in contrast with its potential.

In the framework of a large urban renewal operation, the building was renovated and walls were punched through to open the park onto the city. The Park of the Savèze has become the heart of a neighborhood, accommodating most of its public facilities. It is organized around a connecting thoroughfare flanking the Savèze River and extending the opening of the building.

The river begins in the park via a duct that has collected rainwaters coming from 40 hectares of industrial sites, as well as the flow from upstream creeks. The purification pipes are started upstream in order to bring the water to the ground level of the esplanade and to purify it in a garden equipped with two filters with complementary functions. This garden is the threshold of the park. The water, thus 'renatured', can then continue into the leisure areas before reaching the river bed, making it a more tranquil space. This very natural park offers a haven of peace for the rediscovered biodiversity.

"布列塔尼犁沟"（Brittany Furrow）这个名词，不仅仅用于指代地质断层，还用来指代一栋近乎1千米长100米高的巨大建筑物，该建筑物在1970年代初期建设于南特郊区的圣埃尔布兰。建设时期，规划部门以最小化的方式整治了位于建筑物南面城市边缘地带的一个面积为6公顷的公园。由于受到最小化治理的影响，这个公园少有人光顾，在生物多样性方面乏善可陈，更看不出有什么生态上的发展潜力。在一个大型城市改造方案的规划下，这栋建筑物将被翻新，而且多处墙体将被打通，以便使人们通过公园进入市区。在其大部分公共设施投入使用后，拉萨维泽公园将成为当地的社区中心。公园围绕着拉萨维泽河畔的一条通联性大道展开，扩展了"布列塔尼犁沟"的开放性。

拉萨维泽河发源于公园，最开始是一条流水的小沟。用管道输送进公园的雨水来自邻近面积有四十多公顷的工业建筑和上游山溪。净化管路从溪流上游开始，以便把水输送到地表水平，在一个有两个净水器的花园中使水得到净化。这个花园可说是拉萨维泽公园的枢纽。在那里，被净化的水接下来继续进入休闲区，并最终被注入河流。由此一个更加安静祥和的园区被营造出来。这个环境保持高度自然的公园，为发展生物的多样性提供了天堂般的避难空间。

Master plan of the competition ┊ 总体规划图

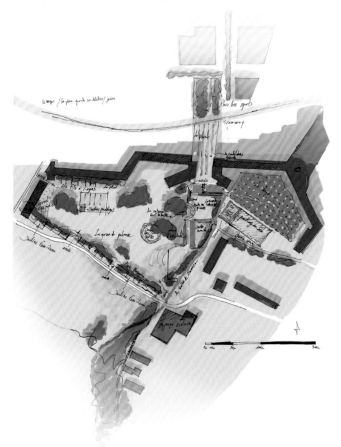

The entrance to the park, fording the filtration garden　01　过滤花园围绕的公园出入口
The route of the water structures the pedestrian paths　02-03　水的流动路径决定了人行步道的组织

Technical cross-section of the filtration garden ┊ 过滤花园的横切面技术图

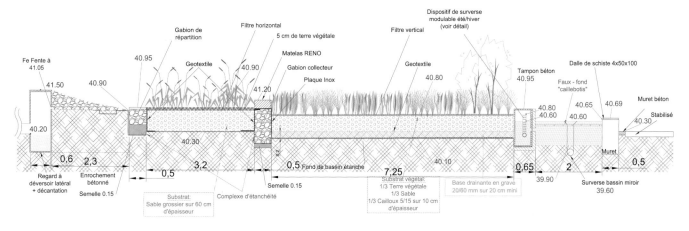

The esplanade and the promenade along the river 04–05　河岸边的休憩场和散步道
The filtration garden at the threshold of the park　06　公园入口的过滤花园

Promenades and play areas along the river 07-09 河岸边的散步道和游乐场
The natural space preserved in tranquility and the 'ecoboxes', 10-12 受到保护的宁静自然空间，其中设置了
island refuges for small fauna   一个个提供小动物栖息的"生态匣"

The Docks' ZAC (comprehensive development zone), a vast, flat territory of 110 hectares beside the Seine, has been set aside for the construction of 3 000 homes bordering the historic center of Saint-Ouen and outside Paris. It will transform the almost monofunctional industrial territory into one fully integrated into the existing city, providing a functional mix and forming part of a wider metropolitan and regional plan that, all going according to plan, will include an eco-neighborhood.

The Saint-Ouen Docks park project must be seen on the triple scale of: the new neighborhood, the Saint-Ouen area, and the Paris metropolitan area.

Agence Ter

Docks Park
码头公园

Location｜地点
Saint-Ouen, France
Completion date｜完工日期
2013
Area｜面积
12 ha
Client｜业主
Sequano Aménagement
Contracted partner｜合作事务所
Agence Ter architectures, Coup d'éclat, Phytorestore, Biotope
Photo credits｜图片版权
Agence Ter (n°03&09), Yang Chen (n°01–02, 04–08)

圣图安市的码头协议发展区位于塞纳河畔，同时临接该城市历史中心和巴黎城门，是一块占地110公顷、用于建造3,000多户住宅的大型街区。此项目将一块位于城市中心、几乎只具备单一工业功能的土地改造成为一个完全和现有城市融合的街区，保证城市功能的混合性，并与大都会和地区性的发展前景相结合，同时也满足了城市建立一个生态街区的强烈愿望。

圣图安码头公园规划的目标在于必须兼顾新街区、圣图安城市以及城市周围大居民区的三重尺度。

The park appears as a succession of strips of varying widths, running parallel to the Seine. Through a series of 'landings' that descend towards the Seine, it will focus on the horizon of the Seine landscape and the Island of the Vannes. Separated by areas of permanent water or flood water, it unfolds in a stepped series of places descending towards the Seine. From the cool terrace at the southern edge of the park succeed a dry valley, then allotments, filtering gardens, a meadow, an oak-lined walk, the main lawn, a lake, the dike walk (Departmental Road 1) and finally the banks of the Seine.

The park is a hydraulic mechanism that manages and purifies the rainwater of the ZAC, using it for its own ends while aiding in the management of the Seine's floodwater. The resulting movement of flux and reflux allows for the installation of very diverse natural environments.

此公园由一系列与塞纳河平行安置、宽窄不一的带状空间所组成，仿佛逐级跌落、朝向塞纳河延伸的台阶布局，使塞纳河和瓦纳岛的水平景观更加突出。整体公园被一些永久性水面空间或可临时容纳洪水的空间所分隔，产生一系列沿着高差微弱的台阶、朝塞纳河方向渐次下降的场所。从位于公园南部边界处的清新露台开始，延展出以下的空间：干涸的斜沟（可作为容纳多量雨水的用途）、共享花园（由居民经营栽种的蔬果园）、过滤花园、草原、橡树林荫小径、大草坪、大水池、散步河堤和最后的塞纳河岸。

此公园也必须扮演水利机制的角色，调控和净化整个发展区的自然降水量，这不仅可以满足发展区自身的需要，还参与了对塞纳河涨潮的调节。这个机制所带来的涨潮和退潮的运动使得多样化的自然生态环境能够在此生成。

Master plan｜总体规划图

01 The receptacle basin of the rainwaters of the ZAC and the rises in the water level of the Seine River　承接雨水和吸收塞纳河洪水的蓄水池

02 The horizon of the Island of the Vannes　水闸岛的天际线

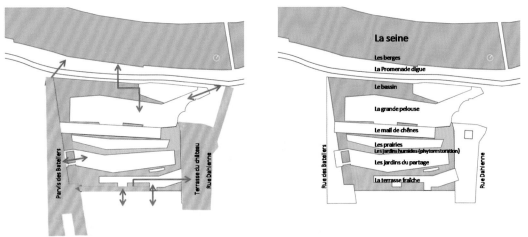

Diagrams of the entrances and the successive entities of the park ｜ 公园出入口个和空间单元分布简图

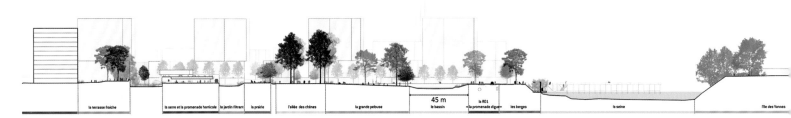

Transverse section from the city park towards the Seine ｜ 从城市街区到塞纳河的公园剖面图

Panoramic view of the park from the hill of games　03　从游戏场小丘看公园的全景
The amphitheatre　04　露天剧场
The playgrounds　05　儿童游戏场

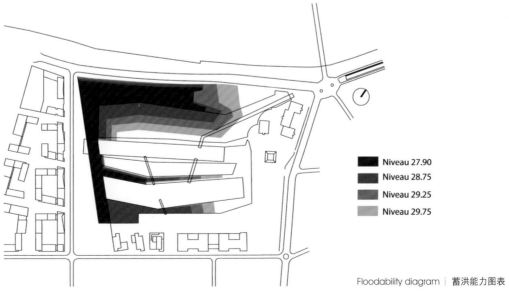

Niveau 27.90
Niveau 28.75
Niveau 29.25
Niveau 29.75

Floodability diagram ┆ 蓄洪能力图表

The shared gardens towards the terrace of the château　06　面向城堡广场的共享花园
The shared gardens　07　共享花园
The waterfalls　08　瀑布
The terrace of the château　09　城堡广场

Agence Ter

Billancourt Park
比扬古公园

Location | 地点
Boulogne-Billancourt, France
Completion date | 完工日期
2014
Area | 面积
7 ha
Client | 业主
SAEM Val de Seine
Contracted partner | 合作事务所
Agence Ter architectures, Biotope
Photo credits | 图片版权
**Agence Ter (n° 01, 05–07),
Yves Marchand (n°02–04, 08)**

The Billancourt park, a major part of the urban renewal project underway on the site of the old Renault factories in Boulogne-Billancourt, fits like a dock harbor basin anchored in the Seine. This long (more than 700 meters), seven-hectare space combines the public park status and that of hydraulic mechanism for the 70 hectares of the new neighborhood in which it stands and whose rainwater it manages.

Its design and its topography give rise to a system of 'islands'. These islands are either accessible spaces or those of a more horticultural nature: the main lawn, the island of cherry trees, and the island of birches. The spaces in between form a collection of places that are sometimes submerged in water, depending to the rain level. A bog, fresh woodlands, wetland meadows, sandy and pebble swales follow on from each other and offer a range of semi-natural environments from very wet to dry. Complementing this role as rainwater manager, which includes keeping the roads free of flooding, the park also serves as a flood basin for the Seine's high waters: during the floods, which occur every 50 years or so, it can be totally submerged…

No superstructure, such as a railing or wall, impairs the perfect visual continuity between the exterior and interior of the park; here, the edges take up the principal of the ha-ha, which is the civic inheritor of the harbour basin.

比扬古公园位于布洛涅-比扬古市的雷诺汽车旧厂房遗址上，是这个正在进行城市街区更新项目里的主导性空间，它临靠着塞纳河，好似河上的船坞。这块占地7公顷的长形土地(纵长700多米)，不仅扮演着城市公园的角色，同时也像是一个水利装置，承担着周围70公顷街区的雨水处理任务。

公园的构图和地形结构显现出一种岛屿系统，"岛屿"指的是一些可供大众使用的空间和以园艺形式呈现的自然景观大草坪、樱桃岛和桦树岛。土地下凹的部分形成了一系列能够承担蓄水功能的空间，它们被水淹没的程度根据下雨量的大小而变化。泥炭沼泽、清凉的林下灌木丛、湿润的草原、沙地斜沟与沙滩依次排开，形成了一系列从潮湿到干燥的各种半自然生态场所。除了起到处理自然雨水和道路积水的作用，公园还起着平衡塞纳河潮汐涨落的作用。按照此规划设计，一旦50年一遇的洪水来临，整个公园将完全被淹没。

为了保持公园内部与外部之间完美的视觉连续性，本方案设计不存在任何界定公园边界的地面护栏或墙体。在这里，公园边界的处理重新采用古代护城河的原则。

- Eaux pluviales, eaux claires issues des toitures et des coeurs d'îlot
- Puits de stockage des eaux de pluies
- Station de relevage
- zone d'infiltration
- Bassin étanche
- Baches de stockage pour l'arrosage
- Passage de l'eau claire par des noues
- Passage de l'eau claire en canalisation
- Station de deshuilage et décanteur des eaux chargées
- Liaison hydraulique des deux parcs

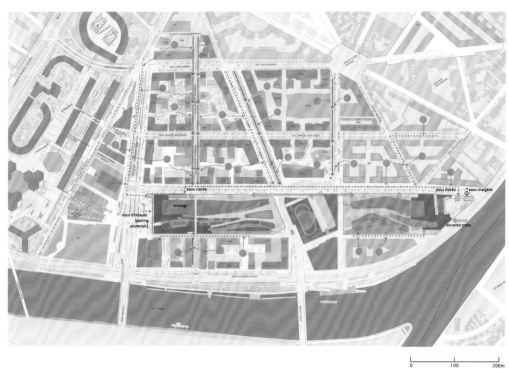

Hydraulic diagram of the ZAC and the park
协议开发区和公园的水管理状况图解

The Island of the Sakuras and the big lawn　01　樱花岛和大草坪
Oblique view of the park　02　公园斜视景观

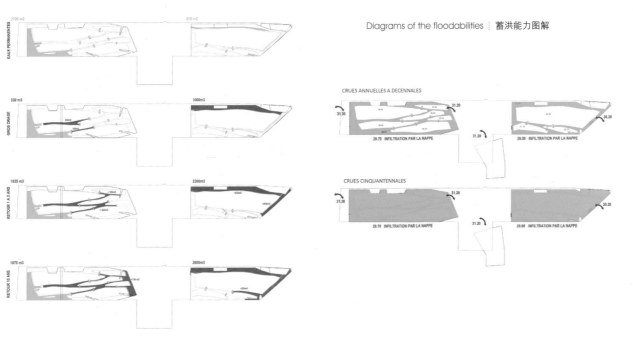

Diagrams of the floodabilities | 蓄洪能力图解

The north bank of the park towards the big lawn　03　朝向大草坪的公园北端水岸
West entrance to the park　04　公园西面出入口
Terrace on the Island of the Sakuras and the caretaker's house　05　樱花岛上的平台和公园管理处建筑

Transverse section of the park | 公园横切图

The springtime arched vault of the Sakuras　06　春季时节路边的樱花如盖
The sandy ditches and the perspective of the big lawn　07　沙土沟和大草坪
The pebble swale　08　砾石浅滩

Florence Mercier Paysagiste

Waterside Park
滨水公园

Location｜地点
Villeneuve-le-Roi, France
Completion date｜完工日期
2014
Area｜面积
0.85 ha
Client｜业主
Ville de Villeneuve-le-Roi
Contracted partner｜合作事务所
ESE
Photo credits｜图片版权
**FMP / Antoine Duhamel (n°01–02, 07–12),
FMP / Florence Mercier (n°03–06)**

A space for relaxation and leisure, Waterside Park is a living place that blends in with the neighborhood's dynamic of renovation and makes use of its position on the banks of the Seine. Creating a dialogue with the river, the project opens a portal to nature in the direction of the river and articulates the transition between urban and natural environments in accommodating varied programs for the inhabitants.

The park in its totality, by its topography in the hollow of a valley, also permits the storing and regulation of the neighborhood's runoff water through a clever working of levels. The vocabulary and the design of 'vegetal islands', the different dikes and pontoons that permit regulating and storing water contribute to the riparian poetics of this new garden, bringing a new dream, an echo the Seine, to the center of the neighborhood. An ecological pond serves educational functions as well as being a place to accommodate birds and animals.

The open-valley form of the whole park gives shape to three pools that have the hydraulic function of slowing down the rainwater of the neighborhood. The pools pour into one another by an overflow system that integrates the dikes. The park itself, by its completely redesigned topography, plays in some way the role of a large retention basin without it being perceived as such, except for the fluctuations of the water level at the hollow of this valley, and for the presence of hydrophyte vegetation.

作为一处放松和休闲空间，滨水公园是一个适合游玩的地方，该处公园融合了邻近地区的创新活力，充分利用了自己处在塞纳河岸的优势。项目在朝向塞纳河的方向修建了一条长廊，参观者可以在长廊上欣赏河流上的风景，体验各种各样的项目，尽情倾吐置身于城市和郊野之间时的感受。

公园总体上位于山谷底部，可以通过一种灵巧的调节手段接受、积蓄邻近地区的雨水，并调节蓄水总量的多少。"菜岛"（Vegetal Islands）一词，指不同类型的木栅和浮桥，可用于调节和存储水量，为这个岸边新公园增添几分诗情画意，并把新梦想、塞纳河的风情带到邻近地区。生态池塘承担教育功能，让小朋友们有机会跟鸟类和动物做紧密接触。

整个公园以开放式山谷的形式存在，谷底三个水池的建造深受这种结构影响。借助水池壁中间安装的系统，水池之间的水可以互相流动。公园本身系参考周边地形条件设计出来的，其中生长着众多水生植物，在一定程度上能行使蓄水基地的作用（除了山谷中水位发生变化）。

Master plan of the park｜公园总体规划图

A very refined work on levels to create an opening of nature in the city　　01-02　试图在城市中营造自然景观的细致整地规划

197

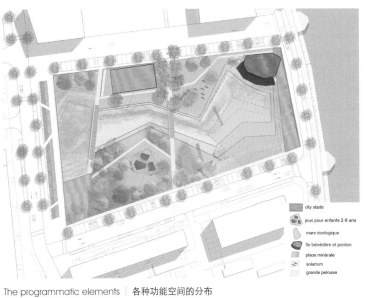

The programmatic elements　各种功能空间的分布

The different strata of vegetation　不同的植被层次

The park, nestled in a hollow, makes possible the storage and regulation of the neighborhood's water

Islands of vegetation, dikes and pontoons create a new imagery in the heart of the garden

03　公园有如一个凹陷的谷地，能够储存和调节整个街区中的水资源

04-06　植物岛、堤道和浮桥在花园中心共同营造出的景象

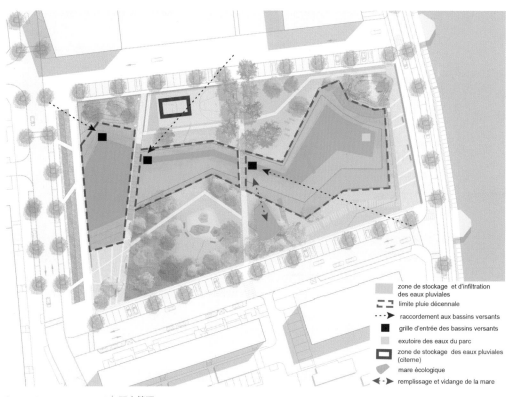

Rainwater management ｜ 雨水管理

- zone de stockage et d'infiltration des eaux pluviales
- limite pluie décennale
- raccordement aux bassins versants
- grille d'entrée des bassins versants
- exutoire des eaux du parc
- zone de stockage des eaux pluviales (citerne)
- mare écologique
- remplissage et vidange de la mare

The extended lines design and stimulate the empty spaces　07-09　延伸的线条设计使得空间显得更加丰富
Ecological and education spaces　10-12　生态和教育空间

Florence Mercier Paysagiste

Ampère Large Garden
安佩尔大花园

Location：地点
Massy, France
Completion date：完工日期
2013
Area：面积
1.5 ha
Client：业主
SEM Massy
Contracted partner：合作事务所
ESE
Photo credits：图片版权
**FMP / Antoine Duhamel (n°01-06),
FMP / Florence Mercier (n°07-11)**

In the center of the new ZAC (comprehensive development zone) of the Atlantis neighborhood, on the site of a former wasteland, the Ampère Large Garden was conceived as a fragment of nature in the heart of the city, offering the inhabitants multiple living spaces and allowing the development of biodiversity.

The project took advantage of several beautiful trees to diversify the existing vegetal strata and multiply the micro-landscapes that enliven the promenade. The work on the soil evokes the stratification of the history of the place. Pleats in the earth were created to uncover layers of soil that suggest the transformations of the site through the course of history: the epoch of the fields of flowers cultivated by the Vilmorin establishments, the trees of the Alstom establishments, and finally the new city neighborhood. Organized around this play of levels, rainwater is collected by a succession of sunken terraces, planted with abundant vegetation that follows the pedestrian axis crossing the park.

The rainwater from both the Ampère esplanade and the spaces with stone construction is channeled in a vast ditch that flanks the large promenade. This ditch is actually composed of a succession of cascading pools, connected by an evacuation tube. At the bottom, a nozzle prevents the upstream pool from becoming too full.

大公园位于在阿特兰蒂斯地区附近的综合发展区中心地域，那里以前曾经是一块荒地。大公园被当成是切入城市中心地带的一块自然"碎片"，为当地居民提供了多种空间享受，并为该处发展生物多样性提供了条件。

该项目利用了既有的一些美丽的树木营造植被分布的层次感，通过设置多种类的小型景观，为广场增添了生机。公园的地质展示使人意识到这里是一个历史悠久的地方。土层上的褶皱用于揭示土壤的分层结构，让人们知道当地曾经历过的沧海桑田。维尔莫机构培植的花圃、阿尔斯通机构和邻近新城区最后推出的树木具有强烈的时代感。借助不同层次游览项目的组织，雨水被收集进一系列种植了丰富植被的洼地。行人可以经这些位于公园轴线上的洼地穿过公园。

从大公园内的石材步道和建筑流下来的雨水，被管道引入一条位于大广场侧面的沟渠。这条沟渠实际上由几个连续的水池构成，通过挖掘出来的管路彼此相连。在水池的底部有一个泄洪出口，以免有时水池里的水会过满。

Master plan | 总体规划图

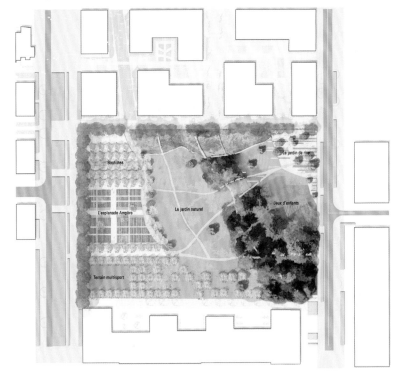

The work on the ground evokes the stratification of the area's history　01　这个特殊的地面层次设计展示了当地的历史阶段发展
Flower gardens and pond vegetation are sunken　02　花圃和植物池塘的地势相对低于
into the main level of the esplanade　　　　　　　　花园大广场的总体地势

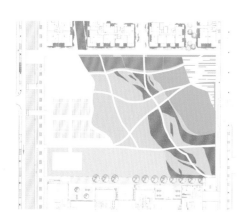

Vegetation stratum 1 – lawn, meadow, undergrowth and flowers

植被层1：草地、草甸、矮树丛和花卉

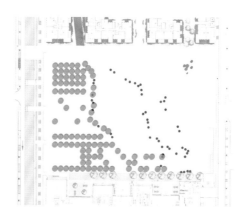

Vegetation stratum 2 – fruit trees, small trees and shrubs

植被层2：果树、小树和灌木

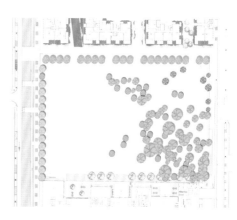

Vegetation stratum 2 – tall trees

植被层3：乔木

Play on the stratification of the soil and of the vegetation　03　各种地面和植物高度的交错搭配
Areas that revitalise the borders　04-06　为边缘地带赋予生机的场所

A succession of ponds that cascade into one another across the garden
The base of the large esplanade accommodates multiple uses for the residents

07-08 互相串联的系列池塘穿越了整个花园
09-11 大广场上的各种设施满足了居民的多种需求

Agence Territoires

Neppert Gardens
内贝尔花园

Location｜地点
Mulhouse, France
Completion date｜完工日期
2015
Area｜面积
0.88 ha
Client｜业主
SERM
Contracted partner｜合作事务所
ANMA, OGI, ON
Photo credits｜图片版权
Nicolas Watelfaugle

At the heart of the buildings, the Neppert gardens offer the residents pedestrian thoroughfares that connect the neighborhood to the rest of the city. This series of themed gardens is designed as a comprehensive system capable of conducting runoff water into the subsoil.

The conversion of the former Lefebvre military barracks in Mulhouse is an opportunity to fundamentally transform the entire Vauban-Neppert neighborhood. The district, which was long considered uninviting, is at the center of an ambitious redevelopment: the preservation of the military architectural heritage, urban renewal, and the construction of hundreds of homes, public spaces, and gardens.

Mulhouse is built on the former bed of the Rhine. The pebbles, accumulated in thick layers by the ancient movements of the river, compose the subsoil of the city. In the gardens, impluviums, or retention basins, provide an intermediary between the surface and subsoil. They collect and conduct the rainwater into the draining subsoil and reveal the geology of the Alsatian plain. The slopes of the garden are all oriented to these impluviums, which end in vast depressions. There, if one looks towards the subsoil, they can see the pebbles of the Rhine, usually buried and invisible. The vegetation of the gardens reflects the nature of this soil, which is particular and familiar to the banks of the Rhine.

这个位于米卢斯市的列斐伏尔旧军营改造项目，为沃邦-内贝尔街区带来深度转变的机会。此街区长久以来被视为人迹鲜少的场所，其改造项目显得雄心勃勃：不仅要达到军事建筑遗产保护的目标，同时也涉及城市更新、一百多户住宅的建造以及公共空间和花园的开创。

内贝尔花园位于建筑群的中心，为邻近街区的居民提供了一条前往城市其他地区的通路，该花园以一系列主题花园的形式呈现，同时也构成一个能够将地面径流导向地底下的大型系统。

米卢斯市位于莱茵河的旧河床上。当地地表有大量的鹅卵石，可以在土壤中造成间隙，将地表水引入地下深层土壤中。花园里的承雨池尤其为地表和地下深层土壤之间的联系创造了有利条件，这一设计带有阿尔萨斯平原的地质地域特色。花园的坡度都朝向承雨池倾斜，在承雨池处形成大洼地，并可借助这些通往地下的"大眼睛"看到一般在地表上看不到的鹅卵石，因为它们通常都被埋藏在地底下。花园中的植物也都是适应莱茵河岸独特土壤环境的种类。

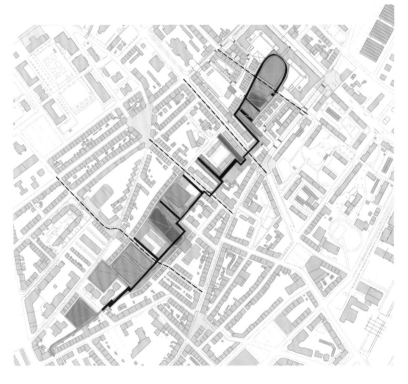

Map of the gardens and the common theme that links them

系列花园以及
整合性元素（红线）的布局

General view of the garden and the impluvium for games　01　花园全景图和游戏用承雨池
The impluviums gather the garden rainwater to infliltrate it　02　收集花园内雨水并将雨水渗透进土壤的承雨池

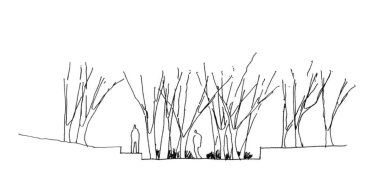

Cross-section of the impluviums ┊ 承雨池的横截面

The 'common theme' crosses the garden from north to south, apportioning the three impluviums　03&05 整合性元素（小径）从南到北穿越花园，连结了三个承雨池

The stonework benches delineate the impluviums and lead the eye towards the draining material	04 石雕长椅界定了承雨池的轮廓，把人们的目光引向排水设施
The strata of pebbles, swept along by the Rhine, compose the Alsacian subsoil and also appear at the bottom of the impluviums	06 被莱茵河顺流冲走而形成的卵石层，在阿尔萨斯地区的下部地层中分布颇多，也出现在承雨池的底部
The rest of the garden provides a contrast because of its emphasis on vegetation	07 花园的其他部分偏重植被的栽种，因此与承水池形成鲜明的对比

Mutabilis Paysage & Urbanisme

Garden of Giants
巨人花园

Location | 地点
Lille, France
Completion date | 完工日期
2009
Area | 面积
3 ha
Client | 业主
Lille Métropole Communauté Urbaine, SAEM Euralille
Contracted partner | 合作事务所
Duncan Lewis Scape Architecture, DVVD, Arc-en-scène, Sogreah, Atelier d'Écologie Urbaine
Photo credits | 图片版权
Max Lerouge (n° 01–03, 05–11), Mutabilis (n°04)

This garden is the setting of the Lille Métropole headquarters, an evocative and exceptional garden created in the neighborhood of Euralille, replacing a large parking lot. Working with a sterile and impermeable terrain stuck between the ring road and bleak architecture, Mutabilis took up the challenge of creating a hybrid garden, one between nature and artifice, that integrated the complexity of its environment. Once the asphalt was stripped away and the few trees in place were preserved, the soil was treated until it had an adequate capacity and receptivity for the installation of rich and contrasting settings.

The water, which is the ever-visible and useful common thread of the garden, flows in a closed circuit. It is guided and directed in various forms: channel delta, ponds, reflecting pools, and aquatic gardens. At the high point of the garden, the water circuit begins, flowing with gravity to the basins of the aquatic garden before returning to the high point via a pump.

In its course, the water feeds certain zones, such as the underbrush of willows that can be flooded by the manual opening of small sluice gates integrated into the channels. Elsewhere in the planted ponds, water is distributed in fine cascades to implement specific environments and vegetation. The presence and dispersion of water throughout the garden creates a lush, rich, and abundant vegetation, sustainably hosting a range of small animals, and ideal for walks and quiet moments.

巨人花园位于欧洲里尔新区、"里尔城市共同体"行政楼脚下，这个极其激发人灵感的独特花园，原本是一个大型停车场。基址被一条环形公路和缺乏生机的建筑所包围，土壤贫瘠而缺乏透水性，穆塔彼利斯接受了设计该项目的挑战在此地设计了一个综合性花园，同时具备自然风情和人工巧制，兼容了周围环境的复杂性。沥青路面的铺设造成了当地树木的减少，土壤受到治理，否则就无法有足够的能力和活力支持丰富多样的建成环境。

水在花园中随处可见，作为花园中常用的组成线索，构成一个封闭的环流，流淌活动于多种方式的存水空间中：呈三角洲状的管路、池塘、水池和水上花园。水循环从花园内的最高点开始，在地球引力的作用下流向名为"清泉花园"的池塘，然后由水泵再抽送到最高点。

在水流动的过程中，水体自然滋润流经的特定区域，例如杨柳的矮树丛。在花园管路系统中设有水闸，当洪水泛滥时，这些闸门将被打开，允许洪水进入矮树丛。此外，在种植有植物的池塘中，为了营造独特的环境和灌溉植被，水流被分流成漂亮的瀑布。水在花园中的径流和分化创造出了大量的繁茂植被，以致园中藏纳了很多适应这种环境的小动物，花园也非常适合散步和养神。

Master plan of the competition: water, the central theme, always visible as well as useful to the garden, circulates in a closed circuit | 总体规划图：总体而言，水随处可见，在花园中起到各种各样的作用，构成一个封闭的环形水道

Water has made possible the transformation of a sterile site into a luxuriant garden　01　水将一块不毛之地变成草木繁茂的花园

The springs garden, the lowest point of the project, is structured by pools　02　位于项目最低点的"清泉花园"由一系列水池所构成

Mist and rivulets in the labyrinth of murmurs 03	薄雾中只闻潺潺水声而不知所踪的的小水渠
According to the specific needs, the water from the rivulets 04-05 can feed certain parts of the garden	小水渠中的水可以根据特定需要灌溉花园的特定部分
The garden's framework in water is readable from some of 06 the buildings of European metropolis of Lille	花园的水文布局可以从欧洲里尔都会街区的某些建筑物上面看得一目了然

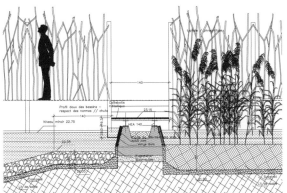

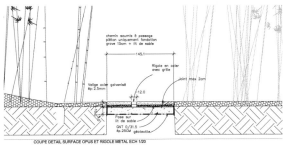

Cross-sections:
details of the paths in the spring garden
and the ground level of the garden

横剖面：
"清泉花园"中的小径和
花园地面细部

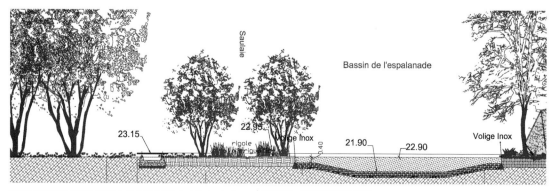

Cross-section: detail of the pond and the coppice of willows ┊ 横剖面：池塘和柳树丛细部

The coppice of willows is furrowed by rivulets of water that feed and structure the ponds 07 一系列小水渠从柳树丛中穿越而过，流入池塘之中，并影响池塘的结构

The animal heads spouting water profit from the omnipresence of water in the project: a pumping system in a closed circuit that feeds them from the pond 08 就近利用项目水资源的动物头形状喷水嘴，与水泵配合从池塘中抽水获得水源，形成一个自给自足的水循环系统

The garden was quickly adopted by the river residents and visitors to the metropolis but also by an aquatic faun that established itself sustainably 09-11 花园很快地被附近居民与来自整个大都会的参观者所频繁使用，同时也引来了许多水生动物，在此长久栖息

Mutabilis Paysage & Urbanisme

Fairy Enclosure
仙子园圃

Location：地点
Paluel, France
Completion date：完工日期
2013
Area：面积
5.6 ha
Client：业主
Commune de Paluel
Contracted partner：合作事务所
COBE Architecture, Berim
Photo credits：图片版权
Luc Boëgly

Located a few hundred meters from the cliffs of Paluel, facing the sea, the Fairy Enclosure is an extension project for the hamlet of Conteville. The site on the clay plateau, next to Paluel's nuclear power plant, is atypical and seems from the outset fairly unattractive…

The project, subjected to rather violent prevailing winds, situated on an almost waterproof soil regularly under mist, makes use of these constraints and finds its own identity and style through them. It is also inspired by vernacular techniques of the region, be they Caux taluses (long mounds topped with tall trees), sunken roads, or thatched-roof buildings.

Concern about water has been at the base of the project's conception. The implementation of a network of drainage ditches that gather all the runoff water of the project is as much a technical tool for water management as it is a natural framework in the neighborhood, allowing the delimitation of spaces and the organization of their boundaries and accesses. The volume of a 100-year rain is stocked in the 'water net', a room in its own right within the Fairy Garden.

Water is always at the heart when it comes to animating, enlivening, and maintaining a garden. A closed circuit of water (ponds, water planches, cisterns…) has been installed, the hydraulic system of which is powered by a wind turbine. Water benefits the gardener, accompanies the visitor, and is the backdrop of stories and tales in the Fairy Garden.

仙子园圃位于距离帕吕埃尔悬崖几百米远的地方，面向大海，是孔特维尔村的一项扩建工程。该项目脚下是一片红土高原，旁边帕吕埃尔核电厂，是一个非典型的花园，人们很容易从一开始就会把这里当成相当不起眼的地方。

该项目所在基址，受到相当猛烈的盛行风影响，坐落在几乎不吸收水分的土地上，而且定期下雾，但也正是这些不利条件使仙子园圃因祸得福，具有了独特的品质和魅力。项目设计也受到当地景观设计手法的影响，比如在斜丘面种植高树、铺筑低于周围高度的道路或搭盖茅草顶的房屋。

对水的关注是项目的核心理念。建造网络化的排水沟渠收集的雨水，是该项目水管理的手段之一。由于基址附近都是荒野，这就使得项目的空间规划、边界确定和道路设定比较自由。园内还有一个被叫作"雨网"的地方，用来存储那种所谓百年不遇的降雨。

水始终处在花园的中心地位，水来到哪里，就让哪里变得生机勃勃、充满生命活力。园中有一个已安装妥当的闭合水循环路径（包括池塘、水池、水塔……），动力来自于风力涡轮发电。水便利了园丁的工作，陪伴着参观者的脚步，也成为那些在仙子园圃流传的故事和童话的背景。

Master plan:
the development promotes a density of residences on the west part of the site in order to create true breathing space

总体规划图：
项目将居住密度集中在基地的西半部，以留出东半部来创造真正开阔的自由空间

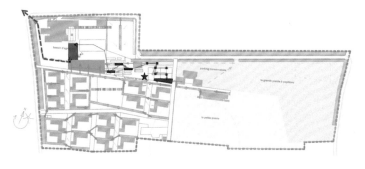 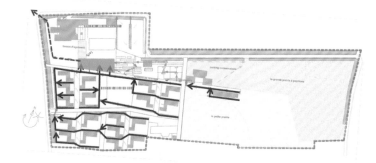

Schematic plan of the different water circuits ┆ 各种水循环的规划图解

The water beds: the water is gravity-fed and feeds the differences spaces of the fairy garden　01　水床：水在重力作用下汇聚而来，并为仙子园圃的各种空间提供灌溉水源

La résille d'eau

Section: between the ornamental pond and water network, the water is stored, highlighted, and utilised for the maintenance of the project | 剖面图：在装饰性池塘和水网络之间，水被储存、用来构成水景，也用于提供项目中各空间的水源需求

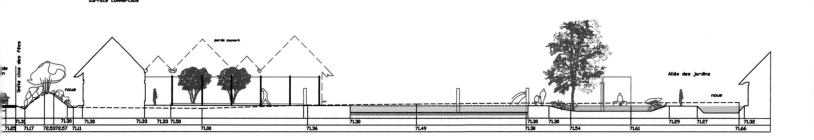

Installed on an almost water-tight clay base, the network of ditches planted with vegetation guides and stores the water and creates the spacial framework of the neighborhood 02 位于几乎不渗水的粘土地基上的绿化沟渠网络，不仅能够引导和储存水流，也为整个街区建立了空间结构

Water is a useful resource in the garden, but is also a tool for enjoyment 03 水是花园中的实用资源，也是创造乐趣的元素

The water management has influenced the toponomy of the areas within the project 04 水管理设施已经影响了项目中各种场所的命名方式

Water structures the useful spaces and the ones for display 05–07 项目的水文系统不仅形成了机能空间的结构，也构成景观空间的布局

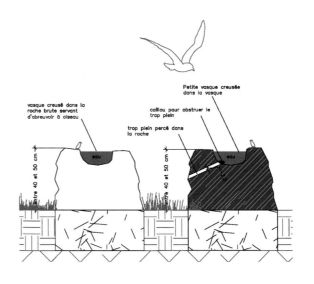

| Rainwater feeds the 'water boulders', little reservoirs for the fauna | 雨水填充"水石槽"，可以为各种动物提供饮用水 |

The ditches planted with vegetation mark the boundary between public and private space 08 种植植物的沟渠标志了公共和私人空间的界限

A wind turbine powers the pump that puts the water in movement in the hydraulic circuit 09 风力发电机组为水泵提供动力，推动水在整个水循环系统中的流动

The network that gathers and stocks the site's water is a space in its own right in the garden 10–11 汇集和储存基地中水体的"水文网络"，成为花园中一个交谊空间的名称

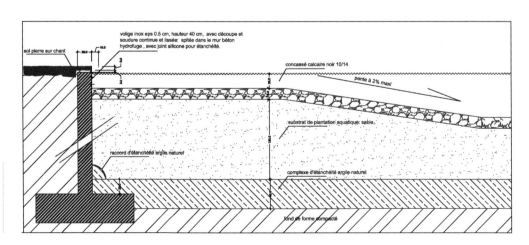

Cross-sections:
Technical details on the fairy pond at the heart of the garden

横截面：
花园中心区域的仙子池塘
详细技术图

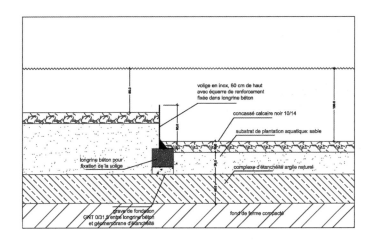
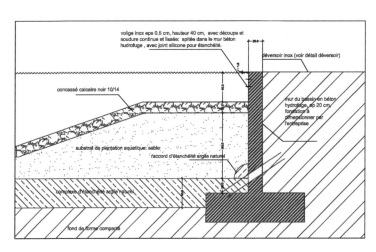

The fairy pond, a water reservoir useful for the water circuit 12 仙子池塘，一个促成水循环的蓄水池
Details on the elements of water management 13–15 水管理机制中一些组成元素的细部大样

Public Rainwater Gardens & Green Spaces

公共雨水花园与绿化空间

L'Anton & Associés

University Campus of Alençon
阿朗松大学校园

Location | 地点
Alençon, France
Completion date | 完工日期
2012
Area | 面积
21 ha
Client | 业主
Rectorat de l'Académie de Caen et Conseil Général de l'Orne
Contracted partner | 合作事务所
Atelier 15, CG61
Photo credits | 图片版权
L'Anton & Associés

The university campus of Alençon was built progressively by the successive aggregation of private, partially public, and public teaching establishments. In 1997, at the time of the construction of new university facilities, the Orne Departmental Council entrusted L'Anton Agency with a study for restructuring the campus. The agency completed the work in 2014, after having created the master plan of the site, defining the urban and architectural parameters for the new buildings, and finishing the development of the exterior spaces.

The climate around Alençon is extreme. The site is windy and the soil is difficult: heavy, impermeable, drying quickly in summer and freezing in winter. For this reason, the water is precious for refreshing the new plantings in spring and summer, but it should not stagnate in winter. The project, from that point of view, is structured by planted and drained ditches, installed as close as possible to the circulations: the pedestrian paths at the foot of the little rest areas, in the separation strips between parking places, and all along the cycling paths or flanking the old departmental route, which has been redeveloped into a bus lane at the heart of the site. The entire project unfolds around the circulation of water, from the building gutters to the temporary creek in the big orchard, which now shelters the site from the winds of the north-west.

With this new water management model in place, the development reinterprets that which humans produce everywhere in this area while they 'make it liveable': collecting, enhancing, and evacuating rainwater at a low-flow rate. The last few decades have often been amnesiac in the face of this technology. And as Jean-Marc L'Anton said: 'There where our ancestors created ditches and mounded hedges, we create pools and wooded bands. But on this campus, we are less constrained by the use of the soil, less prompted to

阿朗松大学校园是逐步建立起来的,因为它先后汇集了私立、半公立和公立的教学机构。1997年,奥恩省议会(the Orne Departmental Council)委托了朗东景观事务所在几栋新校舍竣工之际对大学校园进行重新规划。2014年,事务所完成了这项工作。其中包括校园的总体规划以及相应的城市和建筑规范文本,并且整治了室外空间。

阿朗松地区的气候条件是很恶劣的,该地区不仅多风,而且土壤贫瘠——土质重,不透水,夏季易干燥,冬季结冰。因此,在春天和夏天的新植物种植期,水资源就显得特别重要,但又不能在冬天对水进行储存。从这一点出发,设计规划灌溉和排水沟渠尽可能地接近道路:与人行道相终始,占据休息区的角落,填充停车场划线之间的隔离带,沿自行车道或者旧的省公路侧面分布。旧的省公路现在已经被重新开发为当地核心区域的公交路线。整个项目在我们面前展现出一个水的循环网络,连通了从建筑排水沟到大果园中雨后形成的临时小溪的各个水道,将当地从西北风的淫威下拯救出来。

所有这一切再次表明,人类能实现任何地方的发展,就像在这个地区一样,在水资源较少时,通过收集、积蓄和排放雨水,使自然环境变得"宜居"。过去的几十年里人们经常淡忘这一技术。然而,就像让马克·朗东所说:"那里有我们的祖先挖掘出来的沟渠和栽种的土包似的树篱,现在我们挖掘出池塘,种植出林带。在这个庭园式的地区里,我们很少因土地的状况受到限制,很少采用那些属于'面子工程'的设计,更倾向于营造那些能为人们带来好

Master plan of the project
项目总体规划图

give value to the products of the surface, and more inclined to make other profits, those that have enjoyable structures and that are more easily exploited. But we are just like our ancestors in the search of structures that shelter and protect us, structures in the service of the comfort of users and inhabitants.'

The university campus of Alençon received the Special Jury prize of the Landscapes Victories 2014.

处的设计——让人感到有趣，或者做事情更容易。我们就像我们的祖先一样，寻求能庇护和保护我们的设计，能为使用者和居民带来便利设计。"

阿朗松大学校园这项设计荣获了2014年景观优胜奖的评委会特别奖。

Big ditch along the façade of the new buildings　01–02　沿着新建筑正面设置的大型生态沟渠

Cross-section on the big ditch
and the small squares of the threshold

大型生态沟渠
和入口小广场横截面

Towards the university restaurant　03　大学餐厅方向即景
Hierarchy between the different water collection systems　04-06　不同层次的水收集系统

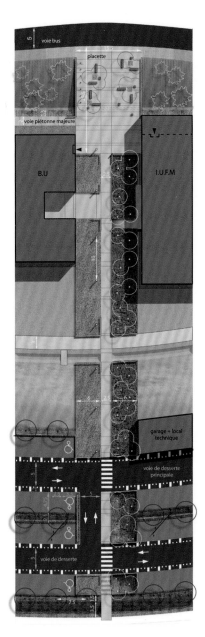

Organizational system
of the different roadways
不同道路的组织体系

Secondary and tertiary ditches 07-08 二级和三级绿化沟渠
Central ditch of the parking lot 09 停车场的中央绿化沟渠

Re-using the existing material: the path was reduced from 6 to 3 meters wide
Permeability of the soil at the foot of the tree and under the seats
The old bench seats are placed on top of gabions

10 现有设施的改造利用：
 道路从 6 米宽被缩减为 3 米宽
11 树下和座椅下地面的渗透性
12–13 金属笼子上面的旧长椅

The landscaping of Saint-Dizier High School was designed to treat and store rainwater, even from the huge floods that occur every fifty years or so. The challenge of this project was to design filtering gardens to treat rainwater that blend into the public spaces, so as not to create any discomfort for the future users of the high school.

Two types of areas were thereby built: inner courtyards at ground level, without ditches or rainwater ponds, and the peripheral areas where the storage and infiltration takes place. This area had to be as compact as possible. Because the rainwater management techniques work with gravity, implementing those solutions according to the waterproof surfaces happened to be more complicated than initially thought, especially around the parking lots.

In order to meet with the levels, the filtering gardens and channels were dug out. Filters are always rectangular strips of grasses planted to obscure their function. Substrate meet the level of surrounding soil. Volume regulation basins and channels are cut in the soil. The goal is not to pretend they are simple natural resurgence. They are breaches to correct the rainwater outflow from buildings. That corrective action, in order to be a good approach to sustainable development, needs to be a voluntary inscription in the soil. Any ecological corrective action is an effort and the students are thereby aware of the presence of these structures.

Phytorestore / Thierry Jacquet

Saint-Dizier High School
圣迪济耶中学

Location：地点
Saint-Dizier, France
Completion date：完工日期
2011
Area：面积
1.02 ha
Client：业主
Conseil Général de la Haute-Marne
Contracted partner：合作事务所
AAT Architecture Jean Philippe Thomas
Photo credits：图片版权
Phytorestore / Thierry Jacquet

圣迪济耶中学的生态景观被设计为可以收集并净化所有雨水，包括五十年一遇的暴雨。该项目的挑战是如何将具有雨水净化功能的过滤花园融入公共空间，并且不对中学将来的规划造成任何影响。

方案设计了两类空间：一类是内院，没有沟渠，也没有雨水收集池；另一类是学校周边空间，以雨水蓄水和渗透工程为主，此类工程需要建造得越密集越好。由于所有的雨水收集设施都以重力作用的原理来运行，实际施工必须依据各种防水收集渠道的地面高差而处理，其复杂程度超乎预期，尤其是在停车场附近。

为了补偿高差，过滤花园和景观斜沟的土方施工挖掘得较深。同时，为了融入景观，过滤池被设计为种植有禾本植物的矩形方阵，填料层与周边土地标高齐平。调蓄水池和斜沟如同在土地上切割出来的深痕，然而其目的不在于造成天然景观的错觉，这些大地上的深痕正是人类为了处理自己的建筑物排放的雨水所做的矫正措施，是迈向可持续发展的补救行动。所有的生态补救行动都是值得为之努力的，这里的中学生们也因为这些明显的措施而对生态意识有了觉醒。

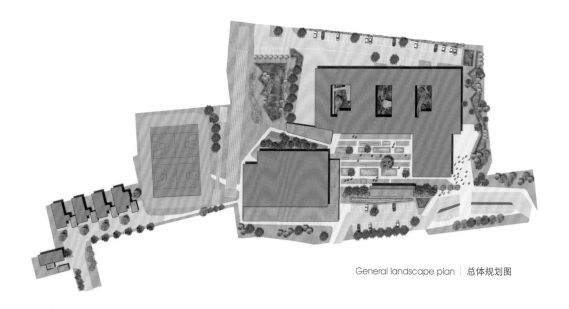

General landscape plan ┊ 总体规划图

Filtering Gardens in the school playground　01-02　学校游戏场的渗水花园
Landscaped infiltration ditches　03　绿化的渗水沟

Sunken buffer ditches	04-06	地势较低的缓冲绿化沟渠
Filters planted with vegetation integrated into the the courtyard	07	与院子融为一体、种有植物的渗水带
Landscaped infiltration strip	08	绿化渗水带

Allain Provost

Citis Business Park
西提斯科技园区公园

Location：地点
Hérouville-Saint-Clair, France
Completion date：完工日期
1987
Area：面积
200 ha
Client：业主
SHEMA
Photo credits：图片版权
SHEMA (n°01), Allain Provost (n°02, 04–06),
Alain Joseph (n°03), Jean-Baptiste Leroux (n°07)

At the gateway of Caen (Hérouville-Saint-Clair), on a wide limestone plateau facing the English Channel, a double approach, both aesthetic and technical, was taken on the gravity-fed and, therefore, natural management of the water of the site. Ditches, canals, ponds, streams, brooks, and rivulets form a north-south network follow the course of an original thin thalweg, where the water from parking lots, roadways, and building rooftops is recuperated.

From one end of an axis to another, one finds in succession: from the entrance, strange geometries of 'tumuli' with five trimmed and sloping sides that give a unique appearance to the place. Then the main feature: a two-hectare rectangular lake divided into three levels, linked by wide staircases of water and simple grassy banks. In the water and on the surrounding land, aquatic plants proliferate, and a sizable and interesting avifauna has established itself. The different gardens of aquatic plants are linked with wooden pontoons that provide a north-south connection. After the lake comes a snake-shaped waterway that feeds the lower pond (created by a dike) with randomly shaped banks.

The aeration and purification of the water are done by all of the flora that colonize the different elements of recuperation. The water ramps of the large lake greatly contribute to the provision of oxygen. There are no particular treatments, such as deoilers. To the north, the water overflowing the lower pond is of a high quality.

在卡昂郊外（埃胡维尔-圣-柯莱尔市镇）朝向英吉利海峡的石灰岩高地，就出现了一个同时注重造型美学和科技特性的项目。方案借助重力的特性来对水进行管理。壕沟、运河、池塘、溪流、小水沟和水渠沿着基地原始的一条隐约可见的谷底线组成了一条水链。

从入口开始，我们便可以看见在中央轴线的两侧排列着一系列修剪为五面锥形的植物体块，成为这个场所的特色。作为基地内的主要元素，一个2公顷面积的长方形池塘处于三个不同的高度，并由大型跌水和造型简单的草坡相连。水中和岸边水生植物繁茂生长，大量而多样的鸟类也栖息其中。各种不同的水生植物园由木质浮桥相连，形成南北走向的通道。湖泊连接着蛇形的水渠为低处（由堤岸围合而成）的不规则池塘提供水源。

通风和水的净化由各种实现不同自然循环作用的植物来承担。大湖中的蓄水极大地有助于邻近地区氧气的供给。当地没有特别的水处理设施，如脱油装置。在朝向正北方的方向，小池塘中的水往往溢流出堤岸，而且具有较高的品质。

Plan　总体规划图：
1. Entrance;　1. 入口；
2. 'Tumuli';　2. 五面锥形植物体块；
3. The lake;　3. 湖泊；
4. The lower pond　4. 低地池塘

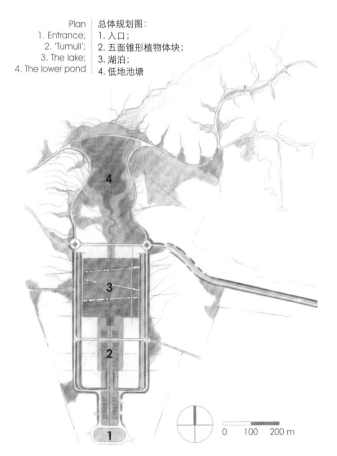

General view of the site	01	项目全景
The big lake fed by rainwater is divided by three grassy dikes that are punctuated by large water stairs	02	以雨水为水源的大湖，中间以台阶式的绿色堤坝作为间隔
View of the lower part of the big lake; these dikes are like large green staircases	03	大湖较低处景象，堤坝看起来就像是大型的青草台阶

A wooden ponton divides the dikes and opens access to the gardens of acquatic plants that have developed up- and downstream of the lower breaks
Spillway between two sections of the waterway
The snake-shaped waterway links the big lake with the lower pond

04&06 将堤坝从中切开的木桥，也是接近水生植物的木桥，这些水生植物沿着切开的堤坝上下游两侧而发展
05 两段湖面之间的泄洪通道
07 连接大湖和低地池塘的蛇形水道

Allain Provost

Renault Technocentre Park
雷诺科技研发中心公园

Location｜地点
Guyancourt, France
Completion date｜完工日期
1992
Area｜面积
160 ha
Client｜业主
Renault
Contracted partner｜合作事务所
Alain Cousseran / Groupe Signes
Photo credits｜图片版权
Allain Provost (n°01, 04–09, 12–15), Renault / Dubois (n°02–03), Jean-Baptiste Leroux (n°10–11)

At the outset, when one considers Renault's history in the area of vegetal environments, no one could have guessed that a few years later one would write: 'The Renault Technocentre is currently one of the largest, one of the most coherent, and most beautiful corporate gardens.' In fact, originally the 500 000 m² of office buildings were to be surrounded by nothing but a sea of 8 000 outdoor parking spaces. The site is a clayey, windy plateau, riddled with rivulets that help in part to feed the fountains of the gardens of Versailles.

For economic reasons, the 2.5 million m³ of rubble from the foundations was used to create wooded hills enclosing the site. The garden is concentrated on the main axis where water is the central theme: lakes, ramps, canals, and ponds succeed one another, largely fed by rainwater from roofs, roads, and car parks. Drained by rivulets, brooks, and ditches, this water is also used in the automatic watering system. It should be noted that as it exits the sizable parking lots, the water passes through deoilers before continuing its course. There is nothing intimate here; priority is given to the large scale: successive zones of water, perennials, grasses, pathways, lawns, and ha-has serving as fences. The lower lake is colonized by aquatic plants, and lotus flowers flourish there.

The water continues on its path through the main building, the 'Ruche' (Hive), to arrive at the central square. Footbridges overlook the water as it plunges via a new chadhar, 'vibrating' water stairs inspired by the Moghul gardens of the 16th century. It flows towards the canal, a 300-meter long fault that seems to be carved into the site. Access to the canal, the site's real garden, can be found along the quays planted with black pines, or at water level, inside the covered galleries.

The galleries cut the canal from east to west, and are bordered with shrubs, perennials, and birch trees—true linear gardens. Such an achievement responds to the current practice of certain companies to improve their image through significant gardens that are more than the ordinary trimmings that typically accompany business 'parks'.

就雷诺工厂过去在植物环境上的文化来看，没有人会想到几年之后有人会写道："雷诺科技研发中心是目前最优质、拥有最漂亮花园的大型办公区之一。"为此，项目主管近乎偏执地邀请了一位景观师来参与方案的构思。事实上，最开始伴随着50公顷办公楼的只是一片容纳8,000个室外停车位的汽车海洋。基地位于一片黏土层的高原之上，多风，流淌着供应一部分凡尔赛花园水池的水渠。

为了节约成本，250万立方米的建筑物基础土方被用来堆积成覆盖着基地、植被繁茂的小山丘。花园集中于主要轴线附近，水流是它的景观主导元素：湖泊、坡道、运河和池塘相连不断，其用水大部分来自于从屋顶、道路和停车场收集的雨水。一些溪流、水渠和滤水沟也被用于植被的自动灌溉。在北侧入口处，有一栋被叫作"前哨"的倾斜的大型建筑，其中设置了四个中庭式花园，建筑物前端依靠着两片平行并列的长条形湖面，它们被种植着小竹子的缓坡隔开，又由波光粼粼的沙塔尔式水坡相连。这里没有任何私密性，最重要的是满足大尺度的要求：序列展开呈带状的水面、多年生植物、禾本植物、小径、草地和界沟。低处的湖面布满了水生植物，荷花盛开。水流继续穿过主要建筑——蜂窝楼，直到中心广场，之后沿着一个新式沙塔尔水坡到达运河，台阶状的水道中水流不断颤动着向前流淌，这个设计系受到16世纪印度莫卧儿式花园的影响。此300米长的运河犹如基地中刻意凿出的断层水带，横向的人行天桥跨越其上。人们可以在运河边地带（基地内的真正花园）沿着种植黑松的河岸行走，或者通过位于低处的、一旁长着水生植物的有凉棚的玻璃长廊前进。

东西向的廊道与运河相交，旁侧种植着矮灌木、多年生植物、桦木，形成真正的线性花园。当前某些企业家希望通过设置大量而重要的花园，而非平庸衬托性景观的做法，来改善其企业形象，但同时又对此种做法能否有效持有相当的疑虑，这个项目的成功为他们提供了最佳的答复。

Cross-section of the upstream lakes 上游湖泊横截面

Overall plan:
1. Entrance;
2. 'Avancée' (Overhang) building and upstream lakes;
3. Plaza;
4. Canal;
5. Car parks

总体规划图：
1. 出入口；
2. "前哨"建筑和上游湖泊；
3. 广场；
4. 运河；
5. 停车场

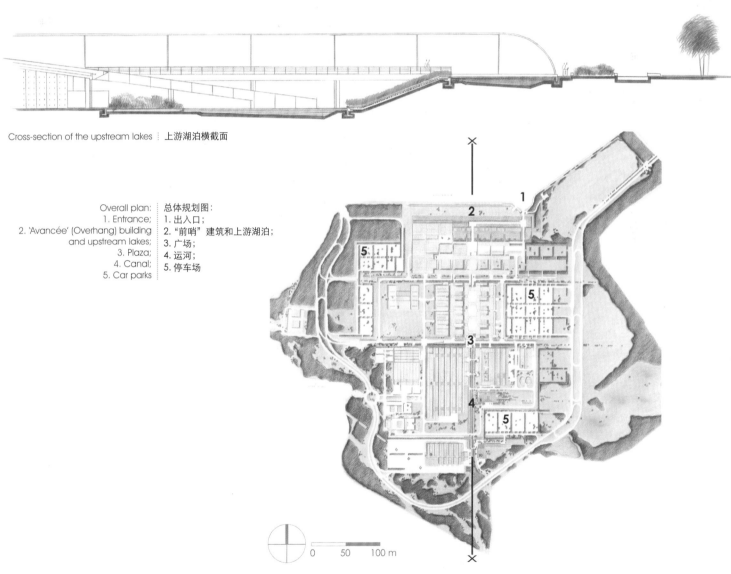

Upstream lake 1 where the water 'goes up' before crossing the site 01 上游1号湖，水流在流经园区前在该处聚集
Patio irrigated by gravity 02 利用重力灌溉的中庭
Northern part of the site with two upstream lakes where the 'Avancée' (Overhang) building extends its 'roots' 03 有两个上游湖泊的项目北部区，"前哨"建筑在此将它的"根"伸入湖泊中

Cross-section of the canal ┆ 运河横截面

The canal, the principal collector of the site's water. Its sections are linked by sloping spillways
The canal is bordered on one side by rough granite walls and on the other side by a mound, the base of which is colonised by acquatic plants
The canal and the entire site are punctuated by covered walkways

04 运河，项目基地中收集水的主要设施，不同河面段落之间由倾斜的水瀑连接
05–06 运河的一边是粗糙的花岗岩墙体，另一边是长满水生植物的坡堤
07–08 整条运河上设置着一道有遮顶的长廊，强调出运河的线条

09 The upstream part of the canal after the water passes across the 'Ruche' (Hive) building

09 经过"蜂窝"大楼处的运河的上游部分

Resurgence of the water　10　水以另一种形态再度现身
Pines emphasize the line of the canal　11　成排的松树加强了运河的笔直线条
Upstream lakes, divided by large water ramps　12-15　上游湖泊分成若干段落，并以宽广的倾斜水瀑相连接，
for maximum oxygenation　　　以便让水产生更高的含氧量

Péna Paysages

Odyssey 2000 Gardens
奥德赛 2000 花园

Location｜地点
Nanterre, France
Completion date｜完工日期
2011
Area｜面积
1.63 ha
Client｜业主
BNP Paribas Immobilier
Photo credits｜图片版权
Christine & Michel Péna

The Odyssey 2000 gardens form part of an essentially industrial site, characterized by a dominance of hard surface and many constraints. The water here, thanks to the river dock and the proximity of the Seine, has a particular look in the light, a riparian vegetation, and river smells. However, it was sought to change its image, which is somewhat negative today: the dock ends in a cul-de-sac in front of the land area and has a color and surface that are far from attractive.

Creating a mainly aquatic park seemed interesting because of the way water calms the spirit. As a mirror, it reflects the moods of the weather that transforms it into a changing surface, sometimes limpid and reflective, sometimes troubled by gusts of wind… The landscape architects imagined it taking on different characters, harnessed and channeled to ensure the transition between the main entrance and the park, a planted lake in front of the restaurants, an ornamental lake to catch the eye in front of the port, a large pontoon giving an effect of stepping back and avoiding immediate contact with the murky waters.

In order to attain the high environmental quality that the site deserved, no ground was waterproofed: grass pavers, stabilized sandy surfaces, and wooden decks ensure a perfect permeability for rain and runoff water. To support the water edges, a variety of hygrophyte vegetation was planted: grasses, reeds, irises, loosestrifes, and other Lysimachea for the ground cover, and ash, birch, willow, oak, pine, and cherry for the trees, with coppiced and row-planted willow for the shrubs.

奥德赛花园位于一块以工业活动为主的基地上，具有诸多限制并且相当缺乏绿化。由于接近塞纳河和船坞，因此水的存在透过特别质量的光线、河岸植被和河流气味，被特别地强调出来。此项目设计目标也在于扭转今日稍显负面的基地形象：码头以死巷的形式在设计地块之前停止，并带着不甚吸引人的色彩与外貌。

由于水具有缓和情绪的功能，在此建造一个以水景为主题的园地成为景观设计构思的重点。镜子般的水面反映着天气的阴晴，呈现出变化多端的面貌，时而澄清明晰，时而被阵阵轻风所撩动。水景形态也呈现多样性格：在主入口与公园之间以沟渠来作为转化空间，在餐厅前面是植物满布的水池，在码头前端则以镜光池面来吸引人们的注意，一座大型浮桥塑造出退缩的效果，避免人们与码头下的混水的直接接触。

为了回应高品质环境认证机构的要求，基地内的所有土地都具有一定的透水性：草地上的铺石、稳固的沙土地面以及木质铺板等，都允许雨水与水流能够很快渗入地下。一系列湿生植物为水岸带来绿化：覆地植物包括禾本植物、芦苇、鸢尾、千屈草和其他的排草属植物；树木包括梣木、杨树、柳树、橡树和樱桃树；灌木则采用低矮和线状的柳树。

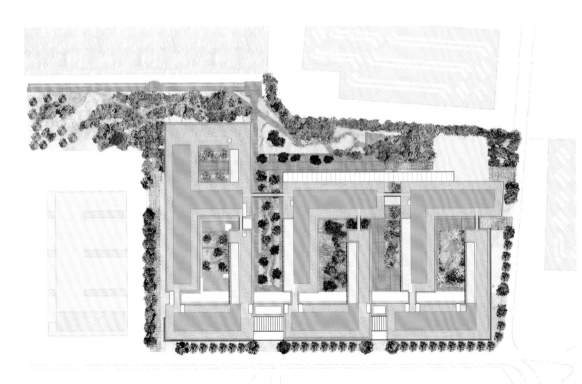

Master plan｜总体规划图

Japanese steps towards the peninsula, 01 朝向餐馆平台对面半岛式花园的
facing the restaurant terraces　　　　　　日式踏脚石
Rythm of the Japanese steps, on the grass and in the water 02 草丛和水中的日式踏脚石分布

Terrace and canal along the harbour 03 码头边的平台和运河
Restaurant terraces around the lake 04-05 沿湖餐馆的露天座位
Above the green theatre 06 绿化剧场的上方即景

Section of the green theatre ：绿化剧场剖面图

Section of the lake ：湖的剖面图

Passage of the iris garden under the offices　07　办公区下面种有鸢尾属植物的水道
Ground through the window above the canal　08　透过办公室窗户所看到的水渠上方步道
Wooden crossing　09　木桥
Patio that leads to the big lake　10　通向大湖的中庭水渠

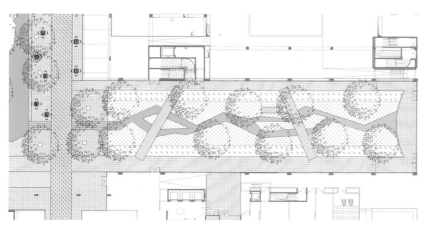

Plan of the patio ：中庭平面图

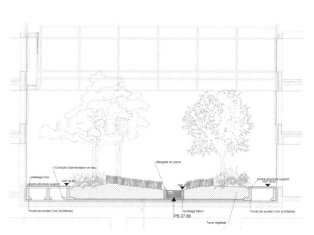

Section of the patio ：中庭剖面图

Bench and pergola　11　长凳和藤架
Floating path above the vegetation　12　植物上方的浮桥
Detail of the bench and pergola　13　长凳和藤架细部
Playful use of Japanese pavement, and paved road　14　日式步道和铺面道路的巧妙结合
Porous path made of grassy paving stones　15　长草的多空隙石头铺面道路

As part of the rehabilitation and extension of shops and amenities, the eco-landscaping designs treat rainwater and wastewater with 'zero emissions' in an urban zone outside of Paris. Treated rainwater is reused for toilets and irrigation of green spaces. Landscape design was also done in order to preserve biodiversity on the site.

The project is interesting because it shows that with a ratio of 20 percent green spaces, it is possible to treat all water on site, without having to link to costly wastewater and rainwater networks.

All the rainwater from parking lots and roads is treated by planted organic filters without using costly traditional techniques and questionable solutions. Filtering gardens, letting the small particles go through, replace a traditional hydrocarbons separator and a degreaser/desander box, which are expensive and polluting.

The site received validation for the 12 HEQ targets, with the filtering gardens taking part in several of those targets, thereby leading to the certification.

Phytorestore / Thierry Jacquet

Saint-Gobain Building Platform
圣戈班建材商场

Location | 地点
Aubervilliers, France
Completion date | 完工日期
2010
Area | 面积
0.46 ha
Client | 业主
Saint-Gobain – Plateforme de Bâtiment
Contracted partner | 合作事务所
Cabinet Architecture Norbert Brail
Photo credits | 图片版权
Phytorestore / Thierry Jacquet

在这个位于巴黎入口的整修与扩建项目中，设计方案通过生态景观的整治来净化处理雨水及污水，达到零排放目标。净化后的雨水被用于冲厕及绿地灌溉。生态景观的设计尤其重视生物多样性的营造。

这个项目的特点在于，在一个绿化面积仅为20%的场地中，其设计方案仍然可以处理全部的雨水及污水，而无需花费大量的投资接管到市政管网。

所有来自道路及停车场的雨水都可通过有机种植过滤系统得到净化，而不需使用效率不高的传统手段。在这个项目中，过滤花园彻底取代了传统的水油分离器及隔油池、除沙器。这些设备不但昂贵，而且对微粒污染物毫无效果。

这个项目得到了法国"高质量环境认证"（HQE）的12项认证，过滤花园在其中好几项的认证中都有所贡献。

Master plan | 总体规划图

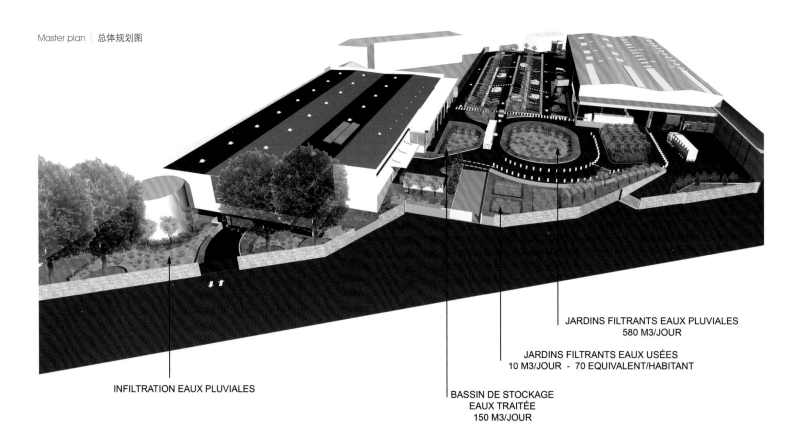

JARDINS FILTRANTS EAUX PLUVIALES
580 M3/JOUR

JARDINS FILTRANTS EAUX USÉES
10 M3/JOUR - 70 EQUIVALENT/HABITANT

BASSIN DE STOCKAGE
EAUX TRAITÉE
150 M3/JOUR

INFILTRATION EAUX PLUVIALES

Large filter in the traffic roundabout 01 交通转盘处的大型过滤花园
Stormwater valleys around buildings 02 建筑物周边的雨水生态沟渠
Entrance to the site 03 基地入口

Filters for all site wastewater 04 废水过滤区
Filtering ditches at the foot of the building 05 建筑物前方的过滤沟渠
Filter in the traffic roundabout 06 交通转盘处的过滤花园
Stormwater filters 07–08 雨水过滤区

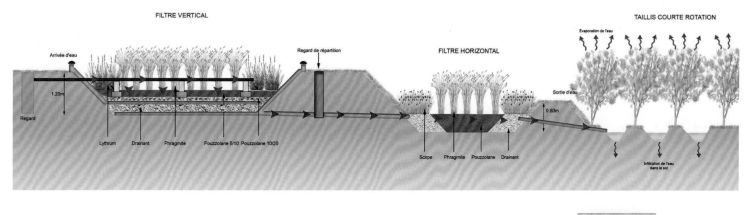
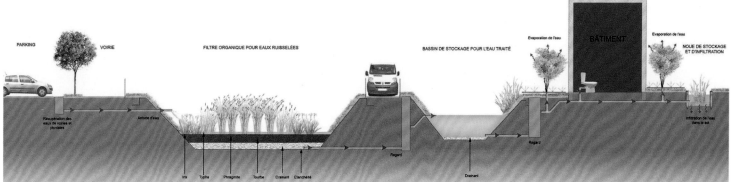

Cross-sections of the filters for rainwater reutilized in the buildings | 雨水过滤花园剖面图，由此得到的净化水可提供建筑物使用

ANNEX
The Designers

附录
事务所和设计师

ATELIER DE L'ÎLE –
BERNARD CAVALIE PAYSAGISTES

3 rue Dagorno 75012 Paris
T +33 1 48 06 22 00
F +33 1 48 06 91 75
bernard.cavalie@atile.fr
www.atile.fr

PP. 136–141

ATELIER VILLES & PAYSAGES
Vincent Roger

170 avenue Thiers 69455 Lyon Cedex 06
T +33 4 37 72 45 01
F +33 4 37 72 27 11
lyon@villesetpaysages.fr
www.villesetpaysages.fr

PP. 92–97, PP. 154–159

AGENCE TER
Olivier Philippe / Henri Bava / Michel Hoessler

18-20 rue du Faubourg du Temple 75011 Paris
T +33 1 43 14 34 00
F +33 1 43 38 13 03
contact@agenceter.com
www.agenceter.com

PP. 184–189, PP. 190–195

ATELIER DE PAYSAGES
BRUEL-DELMAR
Anne-Sylvie Bruel / Christophe Delmar

40 rue Sedaine 75011 Paris
T+ 33 1 47 00 00 51
F+ 33 1 47 00 13 51
contact@brueldelmar.fr
www.brueldelmar.fr

PP. 62–73, PP. 74–87, PP. 116–125

COULON LEBLANC & ASSOCIÉS
Jacques Coulon / Linda Leblanc

40 rue de Fontarabie 75020 Paris
T +33 6 87 52 99 68
T +33 6 82 36 14 59
coulonleblanc@orange.fr
www.coulon-leblanc.fr

PP. 20–23, PP. 88–91

AGENCE TERRITOIRES
Philippe Convercey / Franck Mathé / Étienne Voiriot

22 rue Mégevand 25000 Besançon
T +33 3 81 82 06 66
F +33 3 81 82 08 09
info@territoirespaysagistes.com
territoirespaysagistes.com

PP. 38–43, PP. 142–147, PP. 148–153, PP. 208–211

ATELIER RUELLE
Gérard Pénot

5 rue d'Alsace 75010 Paris
T +33 1 55 04 89 99
F +33 1 55 04 89 69
atelierparis@atelier-ruelle.fr
www.atelier-ruelle.fr

PP. 98–105, PP. 106–109, PP. 110–113, PP. 174–177

HYL HANNETEL & YVER
Pascale Hannetel / Arnaud Yver

90 rue du Chemin Vert 75011 Paris
T +33 1 49 29 93 23
F +33 1 49 29 45 61
paysage@hyl.fr
www.hyl.fr

PP. 24–29

IN SITU
Emmanuel Jalbert

8 quai Saint Vincent 69001 Lyon
T +33 4 72 07 06 24
F +33 1 43 38 13 03
contact@in-situ.fr
www.in-situ.fr

PP. 132–135

MUTABILIS
PAYSAGE & URBANISME
Juliette Bailly-Maître / Ronan Gallais

4 passage Courtois 75011 Paris
T +33 1 43 48 61 33
mutabilis.paysage@wanadoo.fr
www.mutabilis-paysages.com
twitter.com/MutabilisParis

PP. 10–13, PP. 14–19, PP. 212–217, PP. 218–225

ALLAIN PROVOST

5 rue de Naples
T +33 1 39 02 12 55
al.pro@orange.fr

PP. 240–243, PP. 244–249

L'ANTON & ASSOCIÉS
Jean-Marc L'Anton

31 avenue Laplace 94110 Arcueil
T +33 1 49 12 10 90
F +33 1 49 12 10 86
agence.lanton@wanadoo.fr
www.agence.lanton.com

PP. 126–131, PP. 178–183, PP. 228–235

PÉNA PAYSAGES
Christine & Michel Péna

15 rue Jean Fautrier 75013 Paris
T +33 1 45 70 00 80
F +33 1 45 70 72 66
contact@penapaysages.com
www.penapaysages.com

PP. 250–257

TN PLUS
Andras Jambor / Jean-Christophe Nani / Bruno Tanant

30 boulevard Richard Lenoir 75011 Paris
T +33 1 43 55 42 07
agence@tnplus.fr
www.tnplus.fr

PP. 30–33

FLORENCE MERCIER PAYSAGISTE

85 rue Mouffetard 75005 Paris
T +33 1 44 08 80 25
F +33 1 43 31 60 00
secretariat@fmpaysage.fr
www.fmpaysages.fr

PP. 46–51, PP. 52–57, PP. 196–201, PP. 202–207

PHYTORESTORE
Thierry Jacquet

146 boulevard de Charonne 75020 Paris
T +33 1 43 72 38 00
F +33 1 43 72 38 07
info@phytorestore.com
www.phytorestore.com

PP. 58–61, PP. 236–239, PP. 258–261

URBICUS
Jean-Marc Gaulier

3 rue Edme Frémy 78000 Versailles
T +33 1 39 53 14 35
F +33 1 39 49 46 23
axp@urbicus.fr
www.urbicus.fr

PP. 34–37, PP. 162–167, PP. 168–173